Computer Chemistry

Author

Mario Marsili, Ph.D.

Professor
Department of Computer Chemistry
University of L'Aquila
L'Aquila, Italy

CRC Press
Taylor & Francis Group
Boca Raton London New York

CRC Press is an imprint of the
Taylor & Francis Group, an **informa** business

To My Mother
and the KL Eagles

PREFACE

In the last decade we have witnessed a blooming activity in the field of computer applications in chemistry. The reason for this wide acceptance of computer methodologies among chemists may be seen in the particular structure of chemical problems, which can be easily recognized as having strong combinatorial features. It is well known that such problems often resemble solving puzzles in which each stone must be located in one, and only one, proper place to yield a correct final picture. The same happens in chemistry when trying to assemble molecular "fragments", the substructures derived from visual interpretation of spectral data, to form a complete molecule. Similarly, the mental dissection of a molecular structure usually performed by the synthetic chemist to conceive possible synthesis routes is one more classic example where the human brain must tackle monumental combinatorial and permutatorial problems. It was these two main branches of chemical research that stimulated, at the beginning of the 1970s, the birth of the first attempts to combine artificial intelligence and chemistry. We could say that computer chemistry originated in the wish to emulate human chemical thinking within a computer. For this reason, as explained in great depth in the text, computer chemistry must not be regarded as computational chemistry, which is primarily dominated by quantum chemistry. This fact is demonstrated by the history of computer chemistry and its pioneers, the majority of whom were organic chemists. This proves that it was the attempt to reproduce chemical "thinking", and not chemical "computing", that provided the driving force in the primary efforts to compile chemically intelligent computer programs.

The first important schools of computer chemistry were found in illustrious universities in the U.S., Germany, and Japan; this young science had a merely academic character, and many observers just shrugged their shoulders when hearing about "synthesis design programs" or "autodeductive structure elucidation programs". They were somehow annoyed by the possibility that a computer could "think". Computer chemists were considered daydreamers, chemistry hippies not worthy of any serious consideration.

However, the importance of computer chemistry was soon recognized by chemical industry. Its intrinsic potential to enhance laboratory performance was readily made evident, and since then a great deal of funds have been invested for large-scale computerization of industrial chemical research, both in software and hardware.

These last years have definitely seen computer chemistry being accepted even among its previous opponents. Teaching courses are held today in many universities around the world. Learning programming languages has become customary among many chemistry students.

It is further interesting to note how the necessary formulation of chemistry by means of algorithms has been reflected in a clearer view of our conceptual chemical models. The advent of extremely fast computers has cleared the way for the treatment of chemical problems of a complexity unthinkable just 5 years ago. Protein modeling and retrieval of chemical information from data bases containing millions of structural data also have become feasible due to dramatic improvements in hardware architecture. Parallel processors are introducing a revolution in chemical software design and application. Tabletop supercomputers will be available soon, and what appears to be impracticable today will be obvious in a few years. Computer chemistry is evolving at such a speed that any book can seem obsolete if it has to report about the technology. For this reason, this volume is aimed at a conceptual and even philosophical presentation of computer chemistry, enhancing its peculiar psychological aspects; the author has attempted to focus its description on how our human knowledge of chemistry can be transformed into formal schemes, the chemical rules, and then expressed in a form that makes their representation in a computer program possible. This volume is therefore neither a collection of descriptions of the most important computer chemistry software packages nor the exaltation of some specific programs described in more detail than others. It merely attempts to introduce the graduate student, the industrial chemist, the analytical chemist, and the pharmacologist to the world of computer methods in chemical research, which are not alternative but complementary to the currently adopted tools of investigation.

The author has spent more time on the explanation of specific software systems on which he has worked or which he has used frequently. This does not mean that these systems are superior to others that are only cited here: no quality ranking is given for any achievement whatsoever, and judgments are limited strictly to chemical and technical characterizations of the introduced software systems. This book also does not subsititute more specific original literature, but tries to act as a primer for the student approaching computer-assisted methods in chemical research.

Mario Marsili
Rome, Italy
April 1989

THE AUTHOR

Mario Marsili, Ph.D., was born in Rome in 1953. He left his home country at the age of 18 to study chemistry at the Technical University, Munich, Federal Republic of Germany. In 1977 he obtained the "Diplom" degree in chemistry with research work on fast algorithms for the computation of partial atomic charges in molecules based on orbital electronegativity. He earned his Ph.D. at the Technical University 3 years later in the area of computer-assisted synthesis design, where he had expanded the charge calculational models to pi electron systems and first derived bond reactivity functions to be utilized as "deductive" means inside of the synthesis design program EROS, the development of which he contributed to under the leadership of Professor Gasteiger.

He spent one postdoctoral year at the University of Zurich in Switzerland with Professor A. Dreiding, where he worked in the area of molecular graphics and molecular modeling, creating a computerized method for morphological comparison of three-dimensional molecular structures. In 1982 he was appointed Lecturer in Computer Chemistry at the University of Zurich. At the end of 1982 he was called back to Italy by the National Research Council of Italy and joined the team of the Project on Fine Chemistry, directed by Professor L. Caglioti; there he established the first Italian Computer Chemistry research unit. In 1985 he was nominated Assistant Professor of Computer Chemistry at the Rome University "La Sapienza", where he stayed for 3 years. In 1986 he was elected Director of the Strategic Project on Computer Chemistry inside of the National Research Council. At the same time, Italian industry took up the challenge in computer chemistry and an important research project was launched, supported jointly by the Istituto Mobiliare Italiano and 15 Italian chemical and pharmaceutical industries. The project, carried out in the Tecnofarmaci laboratories, was led by Mario Marsili for the scheduled 4 years, ending in the creation of a global molecular modeling system, SUPERNOVA. Currently, he is Professor of Computer Chemistry at the University of L'Aquila and team leader of several industrial research projects in Italy, Germany, and Japan. His actual major fields of interest are molecular modeling and chemometrics.

Dr. Marsili is the author of more than 30 original papers in computer chemistry. He was President of the Ninth International Conference on Computers in Chemical Research and Education, held in Italy in May 1989.

TABLE OF CONTENTS

Chapter 1

INTRODUCTION

I. MAN AND COMPUTERS

Computers have entered most areas of scientific research, industrial production, and educational activities to such an extent that an impact has even been made on the social life, mental attitude, and the psychology of people. Computers can often replace or support many human activities at low costs: cars are assembled by robots; teachers are substituted by computer programs, experienced instructors by simulators. This has occurred because computers are millions of times faster than man. Speed is the name of the game, and speed means competitiveness on the market, low financial investments, and better overall performance. On the other hand, a certain number of disappearing human activities, obsolete and no longer considered profitable, are transformed into new equivalents under a different perspective: the computer perspective. Somebody who in the past manufactured coil springs for wristwatches is almost no longer required, having been replaced by somebody constructing the integrated circuits on which modern watches rely.

Computers have disclosed new frontiers in medicine, improving diagnostic techniques (e.g., imaging in computerized axial tomography). They have caused a real revolution in data management and communication and allow modeling of extremely sophisticated systems like astrophysical events or weather forecasts.

Computers undoubtedly provide a number of astonishing improvements in several sectors of the modern world, but are at the same time the backbone of modern warfare, which has created the most incredible array of annihilating weapons ever (pattern-recognizing "intelligent" missiles, for example). For the single human, this double-faced process of technological evolution has bloomed into a wealth of new professions, all of them connected to computer science, be it theoretical or applied.

Computers are neither good or bad; a knife is neither good nor bad. Each depends on its use. Philosophical fights are raging everywhere on the role of man in a computer-dominated world in which few selected specialists have the knowledge and the power to press strategic buttons on keyboards, and no final solution is expected soon. The question whether human intuition (in other words, the artistic gift, the invention, the intellectual breakthrough) can be replaced by computer simulation, once computers have enough memory and speed to tackle such problems, is indeed a central question and contains even a touch of moral texture.

If a computer simulation based on artificial intelligence systems leads to some unexpected brilliant scientific discovery, is this the merit of the human programmer or of the "thinking" computer?

Chemistry is no exception within the framework of this discussion. The introduction of computer-assisted research techniques into chemistry over the last 15 years has caused a split pattern of reactions among chemists. Whenever computers have been used in a kind of subordinate, secondary, concealed way, they have been accepted as precious and powerful help. This has especially been the case with regard to chemical information and in analytical chemistry. On the contrary, as soon as computers entered an apparent role of equality with the human chemist in solving problems of a more decisional type, exerting a primary, direct influence on man-tailored research strategies and methods, an evident anxiety arose among traditional-minded chemists. Chemists saw (and still see) their leading role as "masters of the art" endangered by an "idiot made of steel". Grown on a serious misunderstanding of the role of computers in chemistry, this attitude in some cases has led to mental rejection of this new technology at the level of its cultural root. On the other hand, enthusiasts are readily found who expect immediate successful results to a variety of difficult problems, believing that "the computer can do everything." They forget that computers still depend primarily on man's performance.

To understand the reasons for a methodology called computer chemistry, to correctly place it among modern research methods, and to detect its benefits and limitations — these points must be discussed in some depth.

II. COMPUTERS IN CHEMISTRY

A. COMPUTATIONAL PROGRAMS

A distinction was postulated above between a direct, or primary, influence of computer action on chemical research and a subordinate, secondary one. Historically this distinction, caused by an independent growth of what is called computer chemistry from other traditional fields of computer applications in chemistry, was rooted in two main facts: the attempt to create computer programs to emulate chemical thinking, and the parallel development of a new, fascinating, and promising branch of computer science, artificial intelligence (AI). AI, which will be discussed later to some extent, is the part of computer science dealing with the computer-generated perception and solution of complex symbol-oriented and semantic problems.

In the early 1970s, chemists were acquainted with a purely numerical use of computers in chemistry. Quantum chemistry and X-ray structure determination were the poles of heaviest exploitation of the fast computational capacity of a computer. In both of these important research fields, the investigator faces such an enormous quantity of bare numbers that their successful treatment would be utterly unfeasible without electronic data processing. The main role of computers in performing these tasks simply consists of managing huge arrays of numbers following a user-implemented, rigid, predetermined prescription. The result of what in a joking manner is termed "number crunching" is in all of these situations a mere numerical result. In other words, the computer delivers a certain number of specific magnitude that interests the user, and the path along which such a number is generated is a one-way road within the codified program. Solving iteratively thousands of Coulomb or exchange integrals and refining Fourier coefficients are examples of such a path. Here the computer follows a fixed scheme of data processing. The final result, for example, could be the energy of some specific electronic state of a given molecule or an array of cartesian coordinates for atoms in a molecule. That is what we expect. The magnitudes of energy and coordinates will change if the investigated substrate is different, but this is obvious. They will also change if a different degree of approximation, refinement, or parameterization is chosen by the user. What does not change is the certainty that some number will come out as the unique result. We might not known in advance what energy value a certain molecule will show at its conformational minimum, but that is the main reason for using a computer: to do the necessary calculations according to user-determined equations which already contain the solution to the problem in all its principles. Due to its advantage in speed, the computer offers a numerical result for final interpretation by man. The program run by the computer contains no alternatives other than to produce quantitative numerical answers of one and the same kind, repetitively, as it has been instructed to do. Truly, there are no alternatives to atomic coordinates for a program that calculates atomic coordinates. The statement "I shall ask the computer to tell me the energy of formation of this molecule" appears to be conceptually and semantically wrong. Justified questioning anticipates the potential existence of an answer; answering demands the *a priori* existence of choice elements among which a suitable answer can be found.

A quantum mechanical program, once implemented according to a particular approach, is geared in a way as to solely calculate a set of numerical quantities, and it has no choice elements on which to exert any kind of deductive evaluation for constructing an answer. Thus, the actual calculation is just a reproduction of the equations contained in the program, substituting real numbers for symbols: no influence is exerted by the computer on the strategic content of the program, on its structure, or on its meaning, and the computer will not be able to change the structures of the equations themselves during execution. Question and answer are like vectors: each has a magnitude *and* a direction in space. The direction determines the difference between a vector and a scalar. Selecting a direction (i.e., including deduction in the formulation of a certain answer by considering the nature of the available choice elements) means adding a qualitative element to a purely quantitative response. Calculating orbital energies cannot produce chemical answers within the conceptual framework just expounded because programs tackling these kinds of computational problems yield scalar numbers (e.g., energies) as results. The direction that we miss in such results, which is nothing less than the general structure of the solution scheme, is called the solution model. In lucky cases of a known theory, this direction is known in advance by the investigator and formulated as a sequence of instructions in a computer program. We can finally assert the following:

Assertion I — *Computational programs in chemistry rely on predefined solution schemes, the models, which are known in their qualitative essence by the user. The output of such programs is a quantitative response, a scalar, for the model under specific, user-given conditions. The generation of such responses follows a rigid, unbranched, and constant data processing mechanism. No strategy evaluation is involved.*

It clearly now appears that computer support in this fashion does not scratch the polished image of any scientist devoting his time to the discovery of fundamental theories or models. He remains master of the situation and welcomes computer aid as a fast and reliable processor of numbers in a kind of subordinate position. In final words, the computer will not teach him anything.

B. SEMANTIC PROGRAMS

What would happen to human psychology and to scientific research if a computer started to deliver qualitative answers, to give strategic advice, to propose models, to change the structure of user input equations, or to emulate chemical reasoning?

To do this, a computer perception of quality must be created. Quality involves comparison; comparison involves rules for judgment; using rules involves the capacity of autonomous acting; acting involves effects; effects involve interpretation and ranking, which finally contribute to the establishment of quality. Quality and quantity together build our response vector, the answer.

Computer chemistry started off right at this point: it provided programs, along with the first blooming achievements and concepts in AI, that were able to help chemists discover strategies. These programs had to be organized flexibly enough to deal with varying mechanisms for making choices. This key term requires the questions addressed to the computer to have, in principle, a manifold set of possible outcomes, which undergo evaluation and ranking.

The intrinsically different response vectors may differ in probability (the magnitude of the vector) and in direction (the quality, the conceptual content of the computer-generated solution, the strategic orientation). Such programs are well suited, in general terms, to provide alternative models, thus enhancing knowledge. That is exactly the complementary (not the opposite) situation to computational programs. The latter apply established models, while the former use general rules (empirical or theoretical), to produce models and ranking strategies. For example, calculating the energy in calories that one needs to move one's arm while playing chess (i.e., to pick up a piece, move it to its new position, and lower the arm again) corresponds to the use of a program belonging to the computational class. However, asking the computer that has been "taught" chess rules to predict all possible sequences of moves leading to checkmate, starting from a user-given initial pattern, is an example of the use of programs of the AI class. Here the process of establishing strategies, predicting countermoves, and ranking sequences of moves according to chance of success is the principal feature of such an autodeductive program.

In computer chemistry, chemical rules are transformed into a program inside a computer, making the electronic device look like it is thinking chemically and therefore turning it into a seeming threat, a cold, stainless steel rival of any human chemist. Computer answers of the following kind are common today, and they make the instinctive repulsion among a few, if not justifiable, at least comprehensible; for example, "Your mass spectrum belongs with 96% probability to a molecule with three chlorine atoms," or "There are 24 different reaction routes within an exothermic range of 0 to 10 kcal/mol that can lead to your desired product; I will draw them for you," or "After interpreting all your spectral data, three molecular structures were found compatible and were generated; here they are," or "You don't have to care for the temperature parameter while running your chemical reactor; adjust the pH to 5.5 instead."

These answers clearly go far beyond those to which chemists had been typically accustomed. They offer direct intervention into operational strategy, as well as tactical realization. They lead to a redesign of a certain experimental setup or to a new, unexpected conceptual insight. Thus, a revised model can be developed. We finally can assert the following:

Assertion II — *Semantic programs are the core of computer chemistry systems. They are tailored to reproduce schemes of human reasoning — in our case, of chemical thinking. They use chemical rules to treat the strategic, decisional kind of problem. They have a primary influence on subsequent methodologies, the establishment of models, the creation of alternatives, and the intelligent interpretation of data in chemical research.*

C. COMPUTER CHEMISTRY AND HUMAN PSYCHOLOGY

The first accomplishment that must be fulfilled is the computer perception and recognition of chemical symbols. Our whole comprehension of chemistry is based on a reiterate confluence of symbols and their chemical content in the human brain, where they are perceived and stored. This process, which takes place over all the years of apprenticeship in chemistry, establishes an automatism that elicits all our chemical knowledge if a visual or phonetic stimulation is conveyed to our cerebral chemical data base. For example, if someone is told the word "benzene", he most likely will visualize in his mind the familiar pictorial symbol for benzene; however, at the same time he will subconsciously correlate to it a number of specific features that he knows are hidden somewhat cryptically in the depiction which certainly belong to benzene as a real chemical entity.

The benzene symbol automatically includes the six hydrogen atoms not drawn explicitly, and the ring inside the hexagon is immediately understood as symbolizing six delocalized π electrons. Even the concept of delocalization is recalled in the brain and is readily formulated as a $(4n + 2)\pi$-electron Hückel rule. This happens at an astonishingly high speed in the human mind. The reason for it is that symbols and their correlated chemical and physical properties are already stored in the brain; they represent our chemical knowledge base. Recalling chemical data (retrieving structural formulas) is a procedure that we do every day while discussing chemistry. A computer does very similar work when used for chemical data retrieval, one of the first applications of computer technology in chemistry. Conceptually, data retrieval is remotely connected to semantic programming, as it generally deals with the matching of input character strings (the name of a molecule, for example) with corresponding strings inside the data base. A relation to truly semantic systems is to be found just in the ability of modern retrieval systems to accept symbols as input, to perform sophisticated logical search and matching operations, and to return the results in an equally sophisticated, symbol-oriented manner. However, no additional original material is generated by the computer during a search session. Autogenous creation of something new must occur by different paths, both in the brain and in computers. Searching for a chemical structure in an array of collected structures stored on some magnetic device can have only one of two possible outcomes: found or not found. In the "not found" situation, the computer cannot augment the data base with the one missing datum because it does not "know" it until an operator supplies the new entry. The unquestionable usefulness of data banks is exemplified by the evident speed in gathering available data as compared to man. The simple psychological experiment of visualizing the benzene symbol and automatically attaching to it all of the chemistry we know (from learning and from practice) highlights the parallelism of our power of perception, our memory, and our retrieving and correlative capabilities with the computer equivalents. These are engineered and emulated inside specific software and deal with a finite set of known elements.

We shall continue this psychological investigation, shifting to problems where new, still unknown elements must be deductively inferred and linked to the previous set. The following argument is an an example of the many possible paradigmatic representations focusing on giving evidence to the differences between man and computer in autogenous creation and manipulation of symbolic elements. It justifies the consistency of inclusion of computer chemistry tools in modern chemical research.

Let us use a different symbol for the representation of benzene, which now will be C_6H_6. This tells us that six carbon and six hydrogen atoms, connected through chemical bonds, form what we call a molecule. Now, in this fictitious experiment, the problem put both to man and computer is to generate all possible structures with the given set of atoms (i.e., generate all isomers of benzene).

The problem is of a semantic/symbol-oriented nature, and according to assertion II its solution requires a number of rules to build the skeleton of the AI procedure. Organic chemistry supplies the rules.

Rule 1. A carbon atom must have four bonds, regardless of its arrangement with connecting partners.
Rule 2. Each hydrogen atom must have one bond connecting it to the next partner.
Rule 3. The molecules must be in a neutral state.
Rule 4. Structures obeying Rules 1 and 2 are valid whether or not they are thermodynamically stable.
Rule 5. No disconnected atoms are allowed.

Disposing of the rules, one can attack the problem of generating as many topological isomers of benzene as possible. Looking at benzene, our fantasy involves the search for a new arrangement of the graphical elements (the lines representing bonds) that constitute the pieces of the game (consider, for example, the analogy to a chess game). The first attempt likely would be to transpose the "localized" double bonds to obtain a new image, as in the case of Dewar benzene (structure **b** below). Another scheme of bond shifting leads to the symmetrical structure **a**, while structure **c**, retaining a hexagonal pattern of carbon atoms, shows one triple and one double bond, with two carbons having more than one hydrogen. If structures **a** and **b** needed only the rearrangement of lines corresponding to double bonds, structure **c** would involve the regrouping of atoms. A major mental combinatoric effort is necessary in abandoning the familiar six-membered ring, which somehow influences inventive flexibility: in the chemist's mind, the hexagon correlates to a flat molecule, a two-dimensional structure. Exploding the 12 available atoms into three dimensions beams to the beautiful structure **d**, prismane.

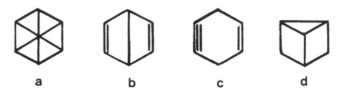

and, going back through a one-by-one structural check, realized that it is the same as

Sooner or later, man's intuition will lead to other images, like open-chain compounds or isomers with five- or four-membered rings in them. The reader may wish to exert himself by finding other elements in the finite set of benzene isomers.

A major difficulty arises when a certain number of isomers have been derived by hand. Suppose that 35 different isomers have been drawn on paper. A 36th is born in the chemist's mind, and in order to validate it he will have to compare the new structure with the other 35. As the mind cannot keep track of so many different images simultaneously, and as they are not perceived and stored in a unique, canonical way, the chemist will in many cases find that the 36th isomer is one that he has generated already. As an example, he might have deduced as the 36th isomer the following open-chain structure,

$$H-C{\equiv}C-CH{=}CH-CH{=}CH_2$$

and, going back through a one-by-one structural check, realized that it is the same as

$$CH_2{=}CH-CH{=}CH-C{\equiv}C-H$$

which he had found long before. The reason is that his mind works on images (symbols), which are remembered not in their abstract, intrinsic nature, but simply as they have been perceived visually; thus, the first linear code given above, once reflected, is at first judged as a different molecule. The brain is not trained for immediate recognition of asymmetrical structures.

The reader interested in knowing how many different structures can be assembled from C_6H_6 and who does not wish to spend the next 6 months doing it without computer help can find them all in the Appendix at the end of this volume. This task takes only a few seconds on a modern mainframe computer.

The human mind seems to be the very best instrument for conceptual breakthroughs, but reveals slowness in exhaustive solution of combinatorial problems. Can the speed at which a computer performs operations be a masked kind of intuition? The great steps in intellectual achievement in man's history were obtained by intuition and not by fast treatment of data according to known rules, as was the case with the benzene isomers. Going from the geocentric concept of the world of the Middle Ages to a heliocentric concept, recognizing the four dimensions of space-time with time being no more absolute, and conceiving particles as waves and waves as particles are examples of the sublime flower of pure intuition, which *breaks rules!* Breaking rules is only in the realm of human thought. Our chemical example proved valuable in understanding the power of a computer in managing data according to rules, but no computer could have such a complete perception of any complex system that it could invent new fundamental rules and, thus, change the boundaries of validity of our rules. This is left to man.

We are now able to confine the role of computers to a well-determined region in chemical research. The computational use of computers requires data to produce data; the use according to AI concepts takes data and rules to produce information, and our minds use intuition to interpret information to finally produce knowledge.

The path between data and information is the area of application of computer chemistry programs.

To end our philosophical digression, we could say that the proper use of knowledge produces wisdom, but this still seems a distant goal for mankind.

Computers can then be instructed to deal with chemical problems where the following hurdles appear to burden human efficiency:

1. An intrinsic difficulty in going from an element n to the next element, $n + 1$, in combinatoric work
2. The creative mind being stained by memories, which are constantly interfering with the new, unborn images we try to bring forth
3. The impossibility of canonical recording of complex structures
4. Danger of redundancy in creation
5. Lack of means to establish the completeness of a finite set of generated elements for a complex system

The reason why computer chemistry diverged from classical computer applications in chemistry (quantum chemistry, physical chemistry, chemical kinetics, X-ray analysis, etc.) and separate journals and conferences were established is rooted in the necessity to deal with formal problems regarding the symbolic perception of molecular structure by computers. Many years were spent generating programs for the perception of rings and aromaticity, for the canonical numbering of atoms in a molecule, for effective user-friendly input and output interfaces, for the recognition and storage of particular substructural features, for the encoding of reaction schemes in reaction data bases, for the fast and compact storage and retrieval of molecular structures, and for the codification of chemical rules. Later, when these basic problems were obliterated, a shift toward a more refined introduction of physicochemical parameters into semantic models, enhancing the chemical quality of computer simulations, took place. Today, due to the enormous speed of mainframe computers (some of them array processors), a greater use of computationally oriented software to feed the semantic, AI-oriented systems with the necessary, more sophisticated data is becoming increasingly popular.

The present stages of evolution show computer chemistry as an established research area constantly propelled by two major mutually supporting thrusts: semantic programs and computational programs.

COMPUTATIONAL PROGRAMS
$\searrow$
$\longrightarrow$ COMPUTER CHEMISTRY
SEMANTIC PROGRAMS
$\nearrow$

III. AREAS OF APPLICATION OF COMPUTER CHEMISTRY METHODS

Imagine an analytical chemist isolating some particular pharmacologically interesting molecule from an animal or plant system and attempting to elucidate its chemical structure. He will use all available modern analytical tools (e.g., high-performance liquid chromatography [HPLC], gas chromatography/mass spectroscopy [GC/MS], infrared spectroscopy

[IR], ^{1}H- or ^{13}C-nuclear magnetic resonance spectroscopy [NMR], elemental analysis, and UV), and if enough substance is available he will then perform some chemical degradation reaction to obtain smaller fragments or target derivatives. All of these investigative techniques provide him with a large batch of raw data that must be interpreted. He knows the rules that link the data (shifts, peak patterns, integrated areas, etc.) to some more or less specific structural elements, the substructures, of the investigated molecules. In an unlucky, difficult case, he may not be able to derive an unambiguous final structure easily, be it due to a possible uncertainty in the molecular formula (MS and elemental analysis do not always guarantee a unique molecular formula; high resolution MS may not be available; etc.) or to the actual combinatorial complexity of assembling the identified substructures. In such a case, the investigator finds an ally in structure elucidation systems: programs for computer generation of molecular structures from spectral and substructural data. These programs belong to the first historic phase of development of computer chemistry tools.

Once the structure of the unknown compound has been elucidated, this information is conveyed to the next laboratory, where pharmacologists, medicinal chemists, and organic chemists work together to find new drugs. The situation can arise where obtaining enough substance for a complete series of pharmacological tests, necessary to evaluate the overall potency of the new drug, becomes cumbersome and expensive because of difficulties in isolation and purification from the natural source. A synthetic approach is consequently decided upon, and by inspection of the target structure some synthesis pathways are proposed by the organic chemist, who proceeds by literature inspection (to find established reaction routes for similar structures) and by intuition. Too often the latter consists of modifications of memorized reactions recalled from the chemical data base in his mind rather than original and innovative contributions. To ensure maximum efficiency in the search for known reactions and to enhance the probability of success in the search for new reaction schemes, he will find it advisable to spend a short time in front of a computer running synthesis design programs. These powerful software systems attempt to model organic reactions, to predict reaction routes retrosynthetically by strategic disconnections of a target compound, and, in a few systems, even to predict the products of an organic reaction from given educts.

Computer and man will cooperate to finally find a suitable way to synthesize a certain amount of the drug in laboratory scale, not focusing so much at this stage on optimization of yield. The drug is tested *in vivo* and *in vitro*, and the pharmacologists become interested in a number of chemical modifications of the current structure to tune its behavior toward a better and lasting biological activity. The design of a first series of analogues of the lead compound includes choosing substitution positions on the parent structure and selecting the type of substituents. Molecular modeling programs provide for a multitude of methodologies to carry out these selections in an optimized manner, and they ensure a means to visualize, manipulate, compare, and describe (by physicochemical parameters) the structures of the analogues.

The analogues will have to be synthesized, and synthesis design systems might be necessary in turn. The analogues are tested extensively, and a number of biological responses are collected (e.g., pharmacological activity, toxicity, time/activity contours, and metabolism). The formation of metabolic products can be simulated by reaction modeling systems in a forward search strategy and their structure inferred by structure elucidation systems, if required. The wish of the investigator now will be to detect a latent link, a structure-activity relationship, between the measured multiple responses and the varying structural features of the analogues. If such a significant mathematical relationship can be found, a second set of more specifically tailored analogues can be postulated by structural modifications which, according to the strategy implied in the structure-activity model, should correlate with increased drug potency, lower toxicity, longer persistence to metabolic breakdown, transport characteristics, and every other drug feature of interest. These kinds of studies, aiming at confirmatory and predictive models, are realized through methods and programs offered by chemometrics. Chemometrics deals with the science of statistics as applied to chemistry. Chemometrics is probably the one direction of computational chemistry that evolved quite independently in the last decade and showed rare connections to the more semantic, strategically operating philosophies described in this book. However, although almost exclusively based on computational programs, chemometrics in its most recent advances seems to gain strategic performances rapidly. Its recurrent application to other systems and the acquisition of semantic outfits rank it among the most prospectively fruitful and promising tools in computer-assisted chemical research. Depending on a variety of circumstances, the chemometrical analysis can be reiterated using pharmacological data measured for the second set of analogues. Suppose that the combined effort of the analytical chemists, the pharmacologists, and the organic chemists seems to converge on a well-defined structure candidate among those tested. It will be necessary at this point to synthesize larger amounts of the substance, and normally this is accomplished in a pilot plant. Optimization of the synthesis procedure suddenly becomes exceedingly important, as it must point to the best conditions for a future scaleup to industrial production and, finally, commercialization of the medicament. In their most recent versions (autodeductive systems, expert systems), chemometrical programs again help the researchers to select those particular experimental parameters which the computer judges to be responsible for the best possible response — in our example, the yield. These selected parameters, the predictors, are then adjusted in practice by the experimenter at predicted trim values corresponding to maximum yield.

This imaginary walk along the several research steps involved in drug design loops back to analytical chemistry when production and quality control actions are requested in an industrial environment to guarantee high standard product quality. Once more, chemometrical programs intervene to sharpen the precision of the collected analytical control data and to ease human interpretation.

In the past, the foundations of structure elucidation systems, synthesis design systems, molecular modeling systems, and related software were established separately. Times were not yet ripe for interdisciplinary overlap, as each field had its own problems finding an inner cultural consolidation, a propositional coherency in the definition of contents and objectives to pursue, and, in many cases, a scientific justification to induce broad acceptance in an initially reluctant chemical audience. Later, the justification was provided by the rising need for more sophisticated drugs, by increasing research times and costs, and by stiff market competition. It must be acknowledged, more to the chemical and pharmaceutical industries around the world than to academic institutions, that an overlap has taken place and that a solid framework of methods in computer chemistry is present today which, although still evolving, successfully operates on a broad spectrum of real problems.

The current architecture of computer chemistry can be represented by Figure 1. Man still rules from the top; at the center is the object of interest, the molecule, around which the various disciplines are positioned, and at the bottom is the computer. All elements of Figure 1 are mutually connected by a conceptual or a real flow of data, by an exchange of information, by some operational necessity coming from, or by a service action addressed to any of the linked elements of the computer chemistry framework.

This chapter has attempted to offer a general introduction to the subject matter of this book, beginning with the mysterious combination of words which forms its title. In the following chapters, the previously mentioned subfields of computer chemistry will be discussed in detail. However, as the laboratory of a computer chemist is a computer and his equipment consists of paper, pencil, and diskettes, a homeopathic amount of knowledge about computer science will be introduced first for readers who are not yet very familiar with computer configurations. Those of you who are comfortable with computer terminology and concepts should proceed to Chapter 3.

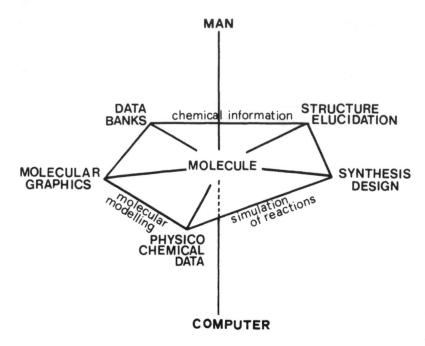

FIGURE 1.   The conceptual framework of computer chemistry. Its main areas of research are positioned around the object of all chemical investigations, the molecule, and are mutually interconnected. Man rules from the top and is supported by the computer (still his subordinate).

Chapter 2

THE COMPUTER AS A LABORATORY

I. HARDWARE

For the computer chemist, computer hardware represents what traditional laboratory equipment represents for the experimentalist — the physical means (and their location) for serving scientific research. Although someone wishing to become a professional computer chemist does not necessarily have to gain a knowledge about computers comparable to a full-time hardware specialist, he certainly will pave his way to a higher final quality of computer chemistry programs if he knows in general terms what can be demanded from modern hardware.

A computer differs from a calculator in self-controlled linking and processing of computational steps, which are contained in one or more programs, called software. A calculator needs a human at all stages of computation. Computers can be divided into two main familes: analog computers and digital computers. Analog computers are machines fed with continuous data, like changing electric currents or other physical time-dependent variables (temperature, light, pressure, etc.) which are emulated internally in analogy to the real physical time-dependent phenomenon. Any input signal to an analog computer can be manipulated and rephrased directly in various fashions by intervention of electronic components of the computer related to a specific mathematical function or operator (multiplication, addition, integration, etc.). Since such a type of computer does not contain logic circuits, as digital computers do, programming is done not at the software level, but through the assembly of electronic parts in a desired sequence. The output is normally some transformed electrical signal whose amplitude can be visualized in several ways, e.g., on the familiar oscilloscope display or on scaled charts. Their use is found primarily in process control: chemical and physical monitoring sensors emit instructional signals to the controlled machine governing its proper functioning. When improper operational conditions are detected, they issue appropriate counteractions or an alarm if they are trespassed.

Analog computers operate in real time and are devoted to the study of dynamic, time-evolving continuous systems. They have no memory and are thus completely neglected in computer chemistry.

Large memory capacity and processing logic are fundamental requirements in scientific computing and simulation of complex systems. They are provided by digital computers, which process discrete electrical impulses encoding numbers, symbols, and operational instructions. The discrete states of these impulses can be represented simply by two states: (1) CURRENT and NO CURRENT, (2) YES and NO, or (3) 1 and 0. The latter representation is a binary representation. Every number and symbol can be transformed into a binary equivalent by binary (base 2) arithmetic. The majority of digital computers are binary machines. The following section will deal specifically with digital machines.

A. ARCHITECTURE OF A COMPUTER

A digital computer is defined as an electronic multiunit system consisting of a central processor unit (CPU), an input unit, and an output unit. The central processor consists of a core memory, a control unit, and a mathematical/logical processor.

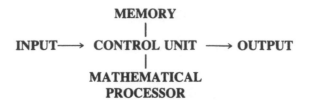

The input and output (I/O) units allow communication between the external world (human, robot, any data storage device) and the central processor. Input devices can be magnetic tapes, disks, keyboards (with visual control through a video terminal), and sensors. In the romantic pioneer era of computers, I/O devices also worked with punched cards and paper ribbons. The atmosphere inside a user's room filled with the "ack-ack" noise of rattling card readers and punchers was more mechanical than electronic. The output unit consists of printers, video terminals, and plotters for direct, human-readable output, whereas fast magnetic or optical alternatives like disks, drums, tapes, and laser-scanned disks allow permanent digitized mass storage of data.

The mathematical/logical unit must be able to manipulate data under the constant supervision of the control unit. Temporarily generated data are stored in accumulators, which are the heart of this unit. In addition, the unit contains the *logic circuitry* responsible for performing the arithmetical and logical operations required by the running programs. Within the core memory, each program instruction is memorized in a codified, machine-dependent numerical form, including all ancillary data. At any time during the data processing, the control unit has direct and fast access to the data contained in the core memory. However, for large calculations, the size of the core memory is in some cases not sufficient to allocate the bulk of incoming data; a memory extension is therefore simulated by modern computers through virtual memory expansion. This technique consists of a dynamic, computer-controlled partitioning and allocation of the requested total amount of memory over core memory *and* fast-access magnetic disks. Thus, programs of a size much larger than the theoretical core memory limit can be processed without forcing the user to cut his program code into subsections small enough to fit the core memory storage boundary. The decision to allocate portions of running programs on virtual memory areas is taken by the control unit, which directs and keeps track of every action inside a computer. The unit reads the current instruction to be addressed to the core memory, interpreting and coordinating all implied operations within the CPU or directed to specific I/O units.

B. BITS, CHIPS, AND MICROPROCESSORS

The elementary quantum of information manipulated by a computer is called a *bit* (from *bi*nary dig*it*), having the two values 1 and 0. This is realized in the hardware by electrical commutators showing two possible states, "open" or "closed", acting like switches or relays. The earliest electrical computers used mechanical relays in a comparable way to establish binary logic. Until 1947, computers were extremely expensive, slow, huge in their physical dimensions, and extremely poor in memory. In 1948, a revolutionary, solid-state electronic device was invented at Bell Laboratories: the transistor. The transistor can be seen as a tiny commutator through which an electric impulse can pass (or not) in a given direction, promptly establishing the necessary connection to binary logic. By replacing vacuum tubes with solid-state technology, the breakthrough toward microelectronics and integrated circuits was achieved. The summarized chronology of such a development is shown here:

Year	No. of elementary components	Vol (m^3)	Price ($)
1955	10,000	20	10^6
1965	10,000	2	10^5
1975	10,000	10^{-7}	10^2
1980	500,000	10^{-7}	10^2

Transistors exhibited a high reliability and a low energy consumption. A trend to min-iaturize computers began at that time and still continues today. The current price collapse of hardware components makes the purchase of a powerful home computer, a personal computer (PC), very attractive. The integration of many transistors and of other electronic elements such as resistors was soon postulated, but for its practical realization more so-phisticated silicon purification and doping techniques had to be developed. Doping means a controlled introduction of trace amounts of alien atoms into the silicon lattice in order to obtain its semiconductorial behavior. The embroidered design of the integrated circuits, i.e., of single quadrangular silicon plates of about 5 mm side length having on their surfaces thousands of transistors, is the result of a repeated overlay of stencils, of masks reproducing one particular scheme of the total circuit. The design is done first on a relatively large scale and then is reduced photographically. Photolithography and other miniaturization techniques make it possible to print many integrated circuits on small slices of single-crystal silicon. They are subdivided into minute plates called chips, each carrying one integrated circuit. The integrated circuit is the strategic elementary unit of modern microelectronics and com-puter technology. The number of components mounted on a single silicon plate has increased exponentially. In 1965, about 10 transistors could be mounted; after 1980, up to 10,000 transistors became the rule.

If one includes resistors, diodes, condensators, and other parts, over 100,000 elements are patched on a single chip. The classification of integrated circuits depends on the number of logical ports, i.e., of functions that can be performed: small-scale integrated circuits (SSI, ca. 10 components), medium-scale integrated circuits (MSI, from 64 to 1,024 components), large-scale integrated circuits (LSI, from 1,024 to 262,144 components), and, recently, very large-scale integrated circuits (VLSI, over 262,144 components).

Chips storing data as 1s or 0s are used to contruct the core memory and the logic circuitry of the CPU of a computer. This last application belongs to the microprocessor's technology. Advanced microprocessors contain all the fundamental parts of a computer CPU and can be programmed in hard-wired form for a broad spectrum of purposes. The specific architecture of a microprocessor determines its speed and the overall system efficiency. Microprocessors are classified according to the number of bits that constitute the basis of the elaborated data. Within one full work cycle, a microprocessor based on an 8-bit architecture can evaluate data that are not larger than the integer number 256 (the highest number obtainable in binary arithmetic with 8 binary digits available); in the same period of time, a 16-bit processor can process data up to an integer of 65,536. However, the number of necessary components increases, too (ca. 100,000).

Eight bits in a row form what is called a *byte*. One byte is enough to translate all symbols of a standard keyboard into a binary machine code. High-performance PCs work with 16-bit microprocessors. In some models, a mathematical coprocessor is linked to the CPU to increase calculation speed. Large computers (mainframe computers) have a 32-bit architec-ture, and the CPUs of some advanced floating point systems (for example, IBM® FPS-164,264) reach the 64-bit level for multiplication and addition operations (vectorial machines, array processors).

The advantage of processors designed on a larger bit basis is rooted in their higher speed of managing a fixed amount of data or, conversely, in processing more data in a given reference time period. They also permit a more compact program structure, with fewer lines of code, due to their own inherent pattern of instructions.

C. MEMORY AND STORAGE
1. Main Memory

In a digital computer, data are memorized in a sequence of bits called cells. All cells (there can be millions) have the same length and are located in a sequence which forms the core memory. The size of a cell is expressed in thousands of bytes: there are 1000 bytes in 1 kilobyte (1 kb), 1000 kb in 1 megabyte (1 Mb), and 1000 Mb in 1 gigabyte (1 Gb).

Inside the core memory, each cell has its own position, called its address. The address can be used to act on a specified section of core memory. It is important to distinguish between the address of a cell and its content (the stored data). Because the address number is represented in binary form, the number of bits needed to trace the location of a particular cell is directly related to the extension of the memory itself. For example, a memory of m cells will have addresses engaging each k bits, with the condition satisfied that $2^k = m$. It follows that the number of bits constituting a cell must be equal to or larger than k.

A processor is connected to the main memory in order to execute the following two actions:

1. The processor can memorize data inside a memory cell while deleting its former content.
2. The processor can retrieve data from a memory cell, leaving its original content unaltered. This action generates a copy of the cell within the processor.

These operations are directed by three components of the processor/memory interface: a store-fetch switch (SFS), a memory-data register (MDR), and memory-address register (MAR). The name "register" is customarily attributed only to cells not belonging to the normal memory cell group. The SFS is a two-bit register indicating the type of interaction to be chosen; the bit string "00" means memorizing, "01" means retrieving data, and "10" leads to an end to interaction. MDR is n bits long and acts as a temporary accumulator of data traveling between memory and processor. MAR has a k-bit capacity and records the address of the cell involved in the interaction. The two interactions can be described schematically through the following steps:

1. Memorizing procedure
   A. The processor sets an address number in the MAR, puts the data in MDR, and switches the SFS to "00".
   B. The memory removes the data from the MDR, transposing them into a cell, and switches the SFS to "10".
2. Retrieval procedure
   A. The processor sets an address number in the MAR and switches the SFS to "01".
   B. The memory makes a copy of the content of the addressed cell, sending the copy to the MDR while the SFS is switched to "10".
   C. The processor retrieves the data from the MDR.

The core memory is the working area inside a computer which contains the programs and the data; it must supply the processor with a flow of instructions. The processor (the CPU), which is the control unit and the mathematical/logical unit, has the task of processing the instructions. Each instruction is split into four steps. The first step determines what kind of operation has to be performed, the second and the third steps determine the memory addresses whose contents are processed, and the last step provides information about the address for final storage.

For this, the CPU uses special registers, selected memory cells attributed to programmed functions. The most important classes of registers are

- Accumulating registers: accumulate the results of calculations
- Memory registers: memorize data taken from or directed to the memory work space
- Address registers: contain a memory address or the address of some peripheral unit
- Instruction addresses: encode the current instruction to be carried out by the processor

The time required to change the data inside a set of registers is called the cycle time.

The main memory can be one of two kinds: a random-access memory (RAM) or a read-only memory (ROM). We previously discussed only the RAM type. Inside a RAM, programs and data will exist only temporarily while the computer is physically running under power supply — in simplest words, as long as it is **on**. Turning off the machine or loading another program into the memory causes the deletion of its former data. Two characteristic features of a RAM are important: (1) data can be inserted and recalled at comparable speeds (read/write memory, or R/W) and (2) the access time to any of the memory cells of a RAM is constant, independent of their physical positions on the chip's board. This is different from other memory devices such as magnetic tapes, where data introduced later on the tape can be recalled after a longer scan time than data introduced at an earlier position.

On the other hand, ROMs can be written at only one single point in time. They are not directly accessible, and their operational information, which is hard wired at the construction stage, cannot be altered. This information regards procedures used in computers to start and initialize the machine (booting), to activate the operating system (OS), and other system or user utility programs.

ROM space in a computer memory therefore cannot be invaded by any user when implementing a program.

2. Auxiliary Memory Devices

Current data are lost inside a core memory when the computer is turned off. Therefore, some means of permanent data storage must be achieved somehow. Devices fulfilling this task are called mass storage devices. The cassette is a frequently used, inexpensive mass storage device which works the same way as a common audiocassette. Instead of sound-generated electrical impulses, the computer-generated discrete electrical signals are stored magnetically on the tape. Since the recorded data are memorized sequentially, like music, in order to access data located, for example, at the end of the tape, one must scan the whole cassette tape. Due to this time-consuming procedure, cassettes are not well suited for fast data access.

Floppy disks are small, flexible disks which are inserted into the disk drives of personal computers. They rotate at a frequency of 300 to 360 rpm. The disk surface consists of invisible concentric tracks on which data are stored magnetically. A floppy disk is a RAM device, as any track can be accessed directly to restore information without having to scan previous tracks. Features that characterize a particular floppy disk are the capability of allowing double-sized recording, the track density (single or double density), the total number of tracks, and its size ($3^1/_2$, $5^1/_4$, or 8 in.).

The mightiest fast-access, permanent storage device is the hard disk, also called the Winchester. It differs from the floppy disk in its physical size and memory capacity (up to several hundred megabytes) and in the fact that it is not removable, being sealed inside its reader case, protected from dust and mechanical damage.

II. SOFTWARE

So far we have briefly described some of the important hardware components of a computer and, using our analogy between a classical chemical laboratory and a computer chemist's laboratory, we could say that the software run by the computer (i.e., the programs written by man) correspond to the various reagents used in "wet chemistry".

It is one of the main tasks of a computer chemist to develop new programs and intelligently use available software on real problems, in the same way as a synthetic organic chemist uses his personal inventiveness and his knowledge about established reaction schemes and reagents to create new synthetic products or novel reaction types. However, in order to offer a more complete description of the subject of this chapter, the computer, we must discuss here solely that particular software, the system software, that realizes all operating functions of a computer, that makes it work. System software is generally a firm-dependent piece of equipment that comes with the installation of a machine; the normal user has made no contribution to it; and it doesn't have any direct access for whatever modification he has in mind. Without this machine-tailored, specific software, a computer would appear more like a big calculator than an intelligent computing machine; the following consideration shall exemplify this point.

The average user of computer facilities writes his own software using a specific programming language chosen from the family of higher level languages. Because many computers (especially micro- and minicomputers) have equivalent CPU structures, all differences in final overall performance can be attributed to the software. It should (1) be written in a language more or less adequately chosen for a particular kind of problem; (2) devote more importance to speed than to structural simplicity (or the reverse); and (3) be generally transportable to other machines or, on the contrary, be geared to run on a single computer model. The program that is to be processed by the CPU, the code, initially compiled in a user-friendly higher level language, is transformed into its binary form within the computer memory. The CPU reads a certain instruction, interprets it, and then performs it. In principle, it is always possible to solve, say, a mathematical equation with a computer by programming it directly using its specific base language, the assembler language. However, in the majority of cases this would prove to be cumbersome and unpractical. One generally should encode the complex operations of the procedure into elementary operations that are executable by the machine; this is done by (1) writing them down in binary code; (2) loading them into the core memory; (3) specifying the starting address of the program; and, finally, (4) providing for the program "start" or "run" instruction. Even more troublesome would appear to be the inclusion of peripherals when storing some data on disk, for example. In this case it is necessary to locate a free block on the disk, making sure that no other information has been stored there previously. Only then can the data be transferred, keeping a record of where they have been stored permanently, in order to be able to restore them later. According to this procedure, the programming work would take so much time that the assistance in speed that a computer can offer the human researcher would be invalidated. For this reason, both operating systems and higher programming languages have been developed.

A. OPERATING SYSTEMS (OS)

An OS consists of several programs, also called routines or jobs, that constitute the interface between user and machine. Through the OS a computer becomes a sophisticated device capable of executing complicated hardware and/or software operations following simple user-issued commands. The monitor is the very heart of an OS. The user interacts with the monitor via a command string interpreter (monitor console routine), which interprets the user's commands, requesting from the OS the activation of other service routines to accomplish the desired operation. For example, a typical command is LOAD; it loads into the core memory a program already codified in base language (machine language). Another common command is RUN; it provides for the execution of the program.

The man/machine interaction and the internal management of the hardware elements are fully controlled by the monitor that allocates core memory requirements and computing time among the various users. It supervises the peripheral devices and steers the data flow in input and output. The monitor also controls the functions of utility programs. These are routines that (1) allow the development and subsequent correction (debugging) of user programs in higher languages and (2) address peripherals for data storage.

The monitor itself resides within the core memory, but must be loaded into it after one has turned on the computer. Only in small computers is the monitor often statically present in the memory and is it active as soon as the power is on, being directly and permanently written on a ROM. The loading of a monitor is called a bootstrap. It is carried out by a bootstrap loader, which initially is conveyed to the central memory or is already available on a ROM. Once the monitor software is read into the memory, the computer is fully operational.

The principal functions of an OS can be summarized as follows:

- Management and control of CPU time
- Management and control of the memory requirement for user programs and for the monitor itself
- Management of the hardware/software interrupts
- Physical management of peripheral devices by means of drivers or device handlers

1. Event-Driven Multiprogramming

Two distinct OS families are defined by single- and multijob monitors. A single-job monitor considers the requests for its hardware/software resources as coming from a single program. This then can be considered as being an integral part of the monitor itself and master of the computer. It can perform all operations except the monitor's destruction. In a multijob architecture, on the contrary, the OS receives different job requests from different sources and must decide about the sequence of actions. For example, two programs might request access to one and the same printer unit or demand use of the CPU; it will be the OS that determines process priority.

In addition, a multijob system can also be a multiuser system; in such a case, it becomes necessary to protect memory areas dedicated to one user from other jobs and other users. Following this, memory and data protection are realized in these systems.

Multijob systems show two criteria in allocating system resources among the running jobs: the priority of the job in event-driven multiprogramming and the partitioning of the available CPU time (time sharing).

The simplest case is found in systems dealing with two jobs, called foreground/background systems. Here the memory accessible by the users' programs is split into two regions. The first region, called the foreground, is occupied by the program with the highest priority (the foreground job); the second region, the background, contains the second program, which has been marked with a lower priority (the background job). Both programs are simultaneously resident in the memory, and the foreground job is processed until it leaves the CPU's control. The execution of the background job then starts promptly and goes on until the foreground job again requests CPU use. This alternating mechanism allows both programs to use the CPU. During the so-called idle time, i.e., while the foreground job does not require CPU intervention, the background job is processed further.

An extension of this operational philosophy to more than two programs leads to the multiprogramming system. Here several jobs compete to obtain control of the system resources. A priority level is assigned to each job. The monitor assigns CPU use to the highest priority job, an operation called priority scheduling.

It is important to recognize that a multijob system implies hardware capable of I/O operations without CPU intervention. This is done, for example, in systems that regulate the data flow with channels, which are CPU-independent microprocessors shifting data directly to or from the core memory, informing the CPU about the termination of a certain data transfer step.

With priority-based systems it is possible for a job with the highest priority requesting only CPU use to exclude all other jobs with lower priority for too much time. Within a time-sharing OS this circumstance is obviated, attributing to each priority-ranked job queueing up for CPU use a predetermined amount of CPU time, the time slice, at the end of which another job is processed. The processing step of any job is terminated if one of following conditions is met:

1.   A time slice has been used up.
2.   The executing job is terminated.
3.   The job necessitates an I/O operation.

An interrupted job (not yet terminated) is relocated within the queue of all other competing jobs. There are different strategies for CPU time assignment: equal time to all jobs, different times to different job ranks, more queues having different priorities, each queue having a different time-slice partitioning, etc.

2. Memory Management

Under a single-job OS, a particular executing job can normally dispose of the whole core memory except the amount allocated to monitor and device drivers. Under a multijob OS, the memory is partitioned into a number of areas. Loading of executable jobs into these areas is a task performed by the OS, which establishes the correspondence between memory area and job. Any active job competes for allocation with other active jobs associated with one and the same memory area. It is possible that an active job momentarily outside CPU control (e.g., during some I/O operation) can be unloaded on a hard disk and kept there for a while (swapping procedure) to allow the introduction of a different job into the memory. Upon reentering the memory, the former job does not have to be loaded on exactly the same memory partition as it had previously occupied, as this area could now be filled by another job. Job reallocation is a very important functional feature of a multijob OS and is realized through a hardware device termed memory management. It is constituted by base registers. The physical address of a certain job memory location is obtained by adding the base register's address to the address that the memory location would show if the job was loaded starting off from the first memory location. Thus, the base register contains the physical starting address of the job.

The same hardware dealing with the reallocation of active jobs and therefore providing for dynamic memory management permits both OS and user jobs to exploit a larger memory than the nominal addressable one.

3. Device Handlers

Device handlers, or drivers, are software packages that when integrated within the OS have control over the physical management of all peripheral units — for example, reading data from or writing data on a specific physical block of a given disk or tape, printing lines of text (records) on some printer or terminal, etc. Every peripheral unit is linked to a specific driver. In general, all of these I/O operations are carried out in asynchronous mode, meaning detached from the CPU, which then can be exploited for other jobs.

The interaction between any device handler and the peripherals is accomplished through hardware interrupts: whenever a data flow between a peripheral unit and the memory becomes active and requires CPU intervention, an interrupt is issued. The system hardware immediately saves the actual machine configuration and starts a driver service routine associated with the interrupt. Upon termination of the I/O procedure a software interrupt is in turn issued and the original configuration is restored; the computer can now continue its job processing at the point where the interrupt occurred. In this situation the monitor takes over the overall process control and performs a priority rescheduling. A system having to survey many peripherals can receive several interrupt issues from different sides. A main scope of duties of an OS is to organize the interrupt calls intelligently, avoiding a severe deterioration of the system's performance, a major loss of computing speed, and the worst scenario, a deletion of data. To achieve this, any interrupt is weighted by a time function according to which the interrupt must be served. If, for example, we consider a data transfer over a 9600-baud line, a single character will be produced every 0.0001 s on average. Within this time the associated interrupt must be served (i.e., the character must be recognized and read by the driver and eventually moved on to its final destination), clearing the area for the arrival of the next character.

Drivers are characterized by the management of peripherals according to the asynchronous techniques just described. Their main inherent advantage consists of making all I/O operations device independent; the user is freed from adapting his I/O operation mode to each and every different physical peripheral device. He just uses general system commands to convey whole ensembles of data, the data buffers, to the various peripherals following one and the same procedure. It is the driver's job to organize the data structure specifically in physical blocks. A physical block is the memory unit of the peripheral (512 characters for a disk and 80 characters for a terminal, for example).

Another important system feature is responsible for easy, user-friendly, and fast data management through peripherals: data management routines allow the user to avoid having to consider how the data are organized on the peripheral unit. Each closed ensemble of data, be it an array of numbers, the source code of a program, or a letter text, is seen as a single entity, a file, which is structured into records. In an I/O step, the data management routines directly read and write such records of a file. The user does not know in detail how these records are placed on the physical and virtual blocks of the peripheral unit.

Files can be distinguished according to access type:

- Sequential-access file — records written (read) sequentially
- Direct-access file — any record on the file directly accessible

The record structure can be one of two types:

- Fixed-length record — all records occupying an equal number of bytes
- Variable-length record

The physical structure of a file on a given device can be one of two kinds:

- Linked — occupies contiguous physical blocks on the device
- Mapped — occupies randomly disposed blocks on the device

4. Higher Level Programming Languages

The most elementary programming language is the assembler language. This language relates a symbolic notation to each instruction of the basic instruction set in order to facilitate the use of the instructions. For example, in some computers the instruction "sum" between two registers (say, R1 and R2) appears as

ADD R1,R2

This instruction is then translated into its binary equivalent

0110000001000010

The translation from symbol notation to binary notation is done by an assembler compiler. The particular features of the assembler language can be summarized as follows:

- There is a one-to-one correspondence between symbols and executable binary commands.
- It is a completely machine-dependent language; its use is only feasible by thorough knowledge of the hardware.

An assembler compiler outputs an object code written in relocable binary code. Relocable means that it is written as if the first instruction occupied the first memory location.

To make the program executable it is necessary to link the object code to the effective memory addresses and to combine different related object codes into a new, unified final object code (relocable) that can be loaded in its entirety into the core memory. This action is performed by a software module called a linker.

Any assembler language, which may be essential in some situations, has several serious shortcomings. It forces the programmer to investigate in depth the architectures of the hardware and of the OS with which he is going to work; an assembler program written for a specific computer will never run on a different machine, and likely not even on the same machine if another OS version is used. Logical steps of the procedure are often difficult to transform to produce complicated routines when formulated in an elementary language. In contrast, in higher level programming languages (e.g., COBOL, PASCAL, FORTRAN, PL1, ALGOL, BASIC, LISP, PROLOG, and C) there is a correspondence between a logical operation and its related semantic instruction. Again, a compiler, which is now the only machine-dependent element, translates the source code into machine-executable binary code. Theoretically, higher level languages should be machine independent, but real life demonstrates that implementation of a program on different machines, under different operating systems, and using different compilers can cause a relevant amount of trouble. Standardization is still far from being achieved today.

Compilers are quite sophisticated software systems capable of detecting errors in the human written code, of eliminating redundancies (i.e., code optimization), and of tracing errors in the structural logic of the program. Obviously, they cannot decide whether the meaning of the current instruction is correct; this judgment can be made only by the programmer following the debugging procedure.

III. BINARY REPRESENTATION OF NUMBERS

Above we discussed the nature of a binary machine, which uses binary quantities to encode numbers, characters, and instructions. Binary means that such quantities can have only two values, 1 or 0. A bit is a binary indicator (flag). To represent decimal numbers with binary indicators it is necessary to use several bits for each decimal number. Four bits are sufficient to generate the numbers 0, 1, 2, 3, 4, 5, 6, 7, 8, 9, 10, 11, 12, 13, 14, and 15. To generate larger numbers, more bits are necessary. The largest number obtainable with n aligned bits in a binary (base 2) system is equal to $2^n - 1$.

Generalizing, we can say that if $x_1, x_2, x_3, \ldots, x_b$ are the numbers of a system with base b, the number $x_1x_2x_3 \ldots x_m$ is equivalent to $x_1b^{m-1} + x_2b^{m-2} + \ldots + x_mb^0$. We can write

	Bit					
Decimal	**3**	**2**	**1**	**0**		
0	0	0	0	0	=	$(2^3 \cdot 0) + (2^2 \cdot 0) + (2^1 \cdot 0) + (2^0 \cdot 0)$
1	0	0	0	1	=	$(2^3 \cdot 0) + (2^2 \cdot 0) + (2^1 \cdot 0) + (2^0 \cdot 1)$
2	0	0	1	0	=	$(2^3 \cdot 0) + (2^2 \cdot 0) + (2^1 \cdot 1) + (2^0 \cdot 0)$
3	0	0	1	1	=	$(2^3 \cdot 0) + (2^2 \cdot 0) + (2^1 \cdot 1) + (2^0 \cdot 1)$
4	0	1	0	0	=	$(2^3 \cdot 0) + (2^2 \cdot 1) + (2^1 \cdot 0) + (2^0 \cdot 0)$
5	0	1	0	1	=	$(2^3 \cdot 0) + (2^2 \cdot 1) + (2^1 \cdot 0) + (2^0 \cdot 1)$
6	0	1	1	0	=	$(2^3 \cdot 0) + (2^2 \cdot 1) + (2^1 \cdot 1) + (2^0 \cdot 0)$
7	0	1	1	1	=	$(2^3 \cdot 0) + (2^2 \cdot 1) + (2^1 \cdot 1) + (2^0 \cdot 1)$
8	1	0	0	0	=	$(2^3 \cdot 1) + (2^2 \cdot 0) + (2^1 \cdot 0) + (2^0 \cdot 0)$
.						
.						
.						
15	1	1	1	1	=	$(2^3 \cdot 1) + (2^2 \cdot 1) + (2^1 \cdot 1) + (2^0 \cdot 1)$

Addition of two binary numbers obeys the rules

$$0 + 0 = 0$$
$$0 + 1 = 1$$
$$1 + 0 = 1$$
$$1 + 1 = 10$$

As an example, summing 3 and 1 in binary mode yields the following result:

$$
\begin{array}{ll}
0\,0\,1\,1 & \textit{(represents 3)} \\
\underline{0\,0\,0\,1} & \textit{(represents 1)} \\
0\,1\,0\,0 & \textit{(represents 4)}
\end{array}
$$

For subtraction the method of the complement is used. In the decimal system, the base 10 complement of a certain integer number n having i digits is given by the difference of $10^i - n$. As an example, the complement of 632 is $1000 - 632 = 368$ ($i = 3$). Therefore, to subtract, for example, 632 from 670, the computer performs the sum of 670 and the complement of 632. The complement of 632 is $1000 - 632 = 368$. Now we can write the sum

$$
\begin{array}{l}
6\,7\,0 \\
\underline{3\,6\,8} \\
0\,3\,8
\end{array}
$$

giving 38 as the expected result. Truncation occurs at the left-most digits, 6 and 3, which are called the most significant digits.

In binary form, the subtraction $4 - 3$ would be expressed as follows within a half byte (four bits):

$$\text{number } 4 = 0100; \text{ number } 3 = 0011$$

The highest binary number obtainable with four bits is obviously 1111, which is $2^4 - 1$. It can be recognized quickly that the complement of 0011, which is $10^{100} - 0011$ (binary notation), is constructed simply by adding 1 to the complement computed for 1111:

$$
\begin{array}{ll}
1\ 1\ 1\ 1 & \\
\underline{0\ 0\ 1\ 1} & \leftarrow \textit{subtraction} \\
1\ 1\ 0\ 0 & \\
\underline{0\ 0\ 0\ 1} & \leftarrow \textit{add 1} \\
1\ 1\ 0\ 1 & \leftarrow \textit{complement}
\end{array}
$$

It follows that a base 2 complement can be generated just by changing 0s to 1s (and vice versa) in the number to be subtracted and then adding 1 to the result. At the same time it is the computer's internal representation of a negative number; this means that 1101 is the binary notation for -3. The subtraction $4 - (3)$ is equivalent to the addition $4 + (-3)$, and we finally can perform the desired calculation:

$$
\begin{array}{ll}
0\ 1\ 0\ 0 & \text{(represents 4)} \\
\underline{1\ 1\ 0\ 1} & \text{(complement of 3; equals } -3) \\
0\ 0\ 0\ 1 &
\end{array}
$$

which gives the expected value 1.

There are other bases suitable to represent numbers — for example, the hexadecimal system. In this system, 16 different symbols are required to form the basic set of numbers: 0, 1, 2, 3, 4, 5, 6, 7, 8, 9, A, B, C, D, E, and F. The hexadecimal notation has the advantage of being more compact than the decimal one. For example, the decimal number 13 is D in hexadecimal notation. For the following more complex real number we have

$$
(2CA.B6)_{16} = [(2 \cdot 16^2) + (12 \cdot 16) + 10 + 16^{-1} + (6 \cdot 16^{-2})]_{10} = 714.7109
$$

To convert binary numbers into hexadecimal numbers the former are divided into groups of four digits, filling up with zeros to the left and right, if necessary. For example,

$$
(1101110001101)_2 = (0001\ 1011\ 1000\ 1101)_2 = (1B8D)_{16}
$$

The reverse operation requires the substitution of a string of four binary digits for each hexadecimal digit.

The binary formalism leads quite naturally to a different level of operations which are connected to human logic reasoning and which are an essential instrument for solving problems linked to the symbol-oriented and semantic nature of our perception of chemistry. These operations are the content of *Boolean algebra*, which throws a bridge to computer-simulated reasoning and therefore to AI. In the next chapter some fundamental concepts of Boolean formalism and some notions in AI will be given in order to prepare the reader to comprehend the similarities between many chemical patterns of thought and the formal and conceptual instruments available in AI.

REFERENCES

1. **Bourne, J.,** *Laboratory Minicomputing,* Academic Press, New York, 1981.
2. **Booth, T. L.,** *Introduction to Computer Engineering Hardware and Software Design,* John Wiley & Sons, New York, 1984.
3. **Ralston and Meek,** *Encyclopedia of Computer Science,* Petrocelli and Charter, 1976.
4. **Horenstein, H. and Tarlin, E.,** *Computerwise,* Vintage Books, 1983.

5. **Andrews, M.,** *Programming Microprocessor Interfaces for Control and Instrumentation,* Prentice-Hall, Englewood Cliffs, NJ, 1982.
6. **Hockney, R. W.,** in *Parallel Computers: Architecture, Programming and Algorithms,* Hockney, R. W. and Jesshope, C. R., Eds., Adam Hilger Ltd., Bristol, U.K., 1981, 25.
7. **Evans, J., Ed.,** *Parallel Processing Systems,* Cambridge University Press, London, 1982.
8. **Wallach, Y.,** *Alternating Sequential/Parallel Processing,* Lectures Note in Computer Science, Vol. 124, Springer-Verlag, Berlin, 1982.

Chapter 3

PROBLEM SOLVING AND ARTIFICIAL INTELLIGENCE

I. BOOLEAN OPERATIONS

The first and most fundamental approach to simulated reasoning is the reproduction of elementary logic, of an inferential (decisional) profile in problem solving. In a laboratory, how often do we hear questions such as "Does this molecule contain a carbonyl group or not?"

TRUE and FALSE, YES and NO, and 1 and 0 are all different symbols for the same logic meaning, which is binary in nature. They are also the logical variables of an algebra ruled by logical operators, the Boolean operators.[1,2,3]

We define the following Boolean operators:

¬	meaning the logical	**NOT**
∨	meaning the logical	**OR**
∧	meaning the logical	**AND**
⊻	meaning the logical	**EXCLUSIVE OR (XOR)**

The following fundamental operations identify the operators' actions on Boolean variables ($1 = true$, $0 = false$):

$$\textbf{NOT:} \quad \neg 1 = 0; \quad \neg 0 = 1$$

$$\textbf{OR:} \quad 1 \vee 0 = 1; \quad 1 \vee 1 = 1$$

$$0 \vee 1 = 1; \quad 0 \vee 0 = 0$$

$$\textbf{AND:} \quad 1 \wedge 0 = 0; \quad 1 \wedge 1 = 1$$

$$0 \wedge 1 = 0; \quad 0 \wedge 0 = 0$$

$$\textbf{EXCLUSIVE OR:} \quad 1 \veebar 0 = 1; \quad 1 \veebar 1 = 0$$

$$0 \veebar 1 = 1; \quad 0 \veebar 0 = 0$$

Graphically we can perceive the meaning of the Boolean operators in defining two sets, $S1$ and $S2$, represented by circles as shown in Figure 1, each containing a finite number of *true* elements which could represent, for example, chlorinated molecules and cyclic molecules, respectively.

A common nonempty subset, $S3$, shall exist, denoted by the area where the two circles intersect. The operator **AND** makes only the elements in $S3$ become *true*, as follows:

$$S1 \quad \textbf{AND} \quad S2 = S3$$

meaning that only the elements that are at the same time members of both $S1$ *and* $S2$ can be members of $S3$. Chemically this means that only chlorinated cyclic molecules are elements of $S3$.

On the contrary, the operator **OR** states that regardless of the set from which an element is chosen it will be included in a new, enlarged set $S4$ of *true* elements; $S4$ will contain all elements of $S1$ and $S2$:

$$S1 \quad \textbf{OR} \quad S2 = S4$$

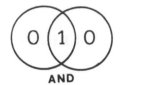
AND

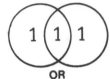
OR

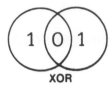
XOR

FIGURE 1.   Two intersecting sets, *S1* and *S2*. Logical operations on *S1* and *S2* (**AND, OR, XOR**) are used to generate special subsets.

S4 includes all molecules on which we are currently operating, the chlorinated species and the cyclic compounds.

The **EXCLUSIVE OR** operator indicates that each element which is a member of either *S1* or *S2*, but not of both *S1* **and** *S2*, can be included in subset *S5:*

$$S1 \quad \textbf{XOR} \quad S2 = S5$$

Here we wish to highlight compounds which are the logical opposites of those in *S3*, regarding as *true* all molecules that have either the feature "chlorine" or "cyclic", but not both features at the same time.

Boolean operators and variables can form more complex equations; for example,

$$1 \lor (0 \land 1) \land \neg (1 \land (1 \lor 1)) =$$
$$1 \lor 0 \land \neg (1 \land 1) =$$
$$1 \land \neg 1 =$$
$$1 \land 0 =$$
$$0$$

II. METHODOLOGY IN PROBLEM SOLVING

The first strategic step toward successfully solving a problem is the clear identification of the problem, which must be recognized with all its peculiarities and in all its intrinsic parametric dimensionality. A clear-cut perception and definition of the goal to be achieved, which makes one feel like walking around the problem, indicates at an early stage which solving tactic is the best to adopt. In some cases more than one solution approach seem reasonable, and only experience will prove which one is the most effective.

The answer to the problem is contained in a model postulated by the chemist; it can be a theoretical, a semiempirical, or a totally empirical model, as will most often be encountered in computer chemistry. The first of several formal actions leading to a computer implementation of the model is the construction of an algorithm representing the logical and mechanistic roles of variables within the model. An algorithm is a procedure for a certain class of problems that guarantees a solution to each problem within a finite period of time or, in the case that such a solution is nonexistent, perceives this nonexistence and informs the user.

An algorithm consists of a branched chain of logical and arithmetic operations representing the framework of the solution method; the key operations of an algorithm are represented schematically with the help of special symbols in a diagram called flowchart.

After a valid algorithm has been established, a computer program must be written and implemented. An algorithm and a program are two different things, the former being a formal solution to a given problem, the latter one of many possible, more or less tricky, and sophisticated translations of the former into machine-executable computer orders. Writing a program can itself be a matter of study, aimed at optimizing execution time and memory requirements, but this will not be treated further here. Instead, we shall focus our attention on the algorithm, on the choice of a particular solution method, on search strategies, and on the related terminology. All this may appear somewhat abstract in this section, but will prove to be of great utility later in the application of algorithmic problem-solving techniques in chemical simulations.

A. DEFINITIONS

We refer here to problem-solving techniques as a group of automated cognitive processes within the range of AI methods.[4,5] The process of problem solving is assumed to evolve within a problem space, Ω, which is an abstract, n-dimensional space containing a finite or infinite number of discrete situations, the problem states μ_k, each defined by a particular set of n parameter values. Goals are special problem states μ^G corresponding to a success, or solution, for a given problem. For example, checkmate is a goal state in a chess game. An initial state, μ^0, is the original problem state, the starting point in the search for a solution. In a chess game it would correspond to all pieces being aligned on the board in their starting positions. A transformation τ is a rule (or a set of rules) that allows the transformation of a given state, μ_k, into another state, μ_i, within Ω. In a chess game the rules controlling the bishop are: run diagonally, no distance limit. Thus, the path from an initial state to a goal can be summarized by the following symbolism:

$$\tau_1(\mu^0) \rightarrow \mu_1, \ \tau_2(\mu_1) \rightarrow \mu_2, \ ..., \ \tau_n(\mu_{n-1}) \rightarrow \mu^G$$

Clearly, we still do not know if the above path from μ^0 to μ^G is the most effective one. It could be just one of a large number of possible transformation sequences leading to μ^G. It is necessary to look for efficient techniques to explore Ω and, consequently, to optimize the path-generating strategies pointing toward a goal.

B. NONHEURISTIC METHODS

1. Random Search

This problem-solving technique acts through a random selection of problem states, jumping from one region of the problem space to another until a goal state is encountered. If a chemist had to synthesize benzoic acid amide (the goal) following this solving philosophy, he would just take any reagent flask he could find in a laboratory without reading its label (the initial state) and add NH_3 to it (the transformation). Sooner or later he would succeed in synthesizing benzoic amide!

Without additional information, such as a possible statistical probability function concerning the density distribution of the problem states in Ω, this approach is too time consuming and unpractical to be the proper choice in complex situations.

2. Algorithmic Methods

Algorithmic methods are based on algorithms, which (as stated earlier) are special procedures, tailored for a particular class of problems, that guarantee a solution if this solution is existent. This is called the principle of completeness, meaning that the algorithm is geared in such a way as not to leave out any one of the possible solutions. The problem is tackled directly and not, as in random search, by means of a blind walk through Ω.

It should be mentioned here that lax use is encountered today for the word "algorithm". In our orthodox sense we mean a mathematical-logical procedure inferring a complete solution for a certain problem and for which proof of this completeness exists. However, it is often misused as a synonym for "procedure", which is only a generic sequence of operational instructions, and sometimes even for "computer program". For example, a hill-climbing procedure[6,7] is a method to individuate the maximum of a multidimensional, continuous, and differentiable function $F(\mathbf{x})$. The method starts from a certain point, $\mathbf{x}^0 = (x_{10}, x_{20}, \ldots, x_{n0})$, in the n-dimensional $F(\mathbf{x})$ space, calculates the maximal gradient $F' = \Delta F(\mathbf{x}^0)/\Delta \mathbf{x}$ around $\mathbf{x}^0$, walks along the steepest direction uphill (to higher function values) until a second point, $\mathbf{x}^1 = (x_{11}, x_{21}, \ldots, x_{n1})$, is reached, recalculates the gradient, and repeats the whole procedure until a point $\mathbf{x}^M$ close to $\mathbf{x}^0$ is found found where $F' = 0$. The initial state is crucial in the described method: if it is located beside a ridge (or a relative maximum), the procedure stops on top of it. It follows that in principle the real absolute maximum may not be found if the conditions are unfavorable; because of this we do not call the hill-climbing technique in the presented form an "algorithm" for the evaluation of functional maxima. It would turn into one if some strategy could be added to perceive relative maxima and force the continuation of the search until a goal is obtained.

C. HEURISTIC METHODS

The term "heuristic" is rooted in the Greek work "heuriskein", meaning "to discover" or "to find". Consequently, heuristic procedures are a computer simulation and programming philosophy suited to finding a problem solution exploiting any empirical strategy, trick, or shortcut by which the computer acquires knowledge of the structure of the problem space beyond its pure abstract definition. Empirical rules, human experiences, or computer self-generated knowledge (i.e., the learning machine) are implemented to optimize the search for the goals. The heuristic methodology does not, however, guarantee (in contrast to the algorithmic approach) the discovery of all possible goals in an absolute sense.

There are statistical methods of heuristic programming in which the dimensionality n of Ω, also evident in the transformation operators τ, is augmented by additional descriptors that relate some peculiar features of the initial states to similar characteristic features of the goals. A statistical search technique trying to establish such links and using them while proceeding along the search path by association, analogy, and similarity of problem states is a heuristic statistical search technique.

Another approach, which plays an important role in computer chemistry, is the generate and test method.[8] Its aim is to generate the goal states directly. Here we do not have a stroll by different μ_k that might lead to the detection of the goal(s) μ^G within the boundaries of Ω. Instead, a solution space Φ is spanned; each and every directly generated μ_k, now called a candidate, is tested to prove its validity as a goal μ^G. In other words, looking for preexistent goals through a stepwise walk within Ω is very different from generating the goals themselves. In the former case their characteristic features are known in advance; in the latter case their parametric structure is unknown. In a chess game we do not explicitly know all of the situations we call checkmate, but we can define their general parametric structure formally as "a given state in which any move of the king from n to $n + 1$ brings about his annihilation". This is a necessary and sufficient condition to discriminate such a goal state when encountered from other situations on the chess board, which are not goals, but only ordinary problem states. However, in interpreting a series of spectra of an unknown compound we do not know in advance what kind of molecule will confront us; what must be created directly, then, are the molecular structures which, being the only kind of problem states of Φ (not of Ω), immediately become the potential goals (the candidates) that must be evaluated further to establish their spectral compatibility. A checkmate situation is recognized as such without error, but it cannot be guaranteed that a computer-generated molecular structure really corresponds to the compound on which the spectral measurements have been performed. An additional test is necessary, which can be a computer-controlled test (by means of AI techniques) or an experimental test controlled by man. We shall deal with this side of the problem in Chapters 6 and 7.

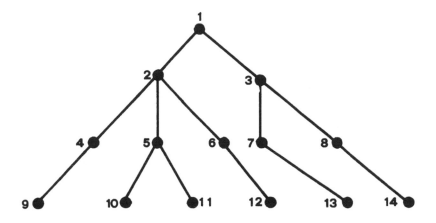

FIGURE 2. Graphical representation of a tree. A tree is a nonmetric entity, and the length of the edges is meaningless. Only the connection pattern between the nodes describes the tree.

1. Trees and Graphs

One of the dominant aspects of AI programming is the use of procedures for the generation of a tree of logical possibilities, i.e., of problem states compatible with the dimensionality, the parametric, and the rule-given requirements included in the AI simulation.[5,9] The expression tree is used because the walk through Ω, going from problem state to problem state, is reminescent of the branches of a tree, which all root in a common stem.

A tree is generated from a starting node, or vertex, corresponding to μ^0. The starting node is transformed by the operators τ to generate other nodes on the tree, also called subgoals. The paths connecting any two nodes are called edges. A subgoal can become the parent of other subgoals (see Figure 2). As soon as a given subgoal coincides with the definition criteria of a goal, terminating conditions for this path on the tree become active.

Each node in a tree descends directly, along a unique path from the initial node. A node generated in a tree has never been generated before.

In a graph, on the contrary, a specific node may be accessed along different paths, meaning that one and the same subgoal in principle can be generated earlier or later, depending on which path is taken. It also follows immediately that some paths starting from a given μ_k loop back to their point of origin.

Figure 3 demonstrates that to reach the goal node $\mu(12)$ the walk on the graph must pass nodes $\mu(2)$ and $\mu(6)$; another successful path is realized with $\{\mu(1), \mu(3), \mu(6), \mu(12)\}$. Generation of $\mu(6)$ is achieved via $\mu(2)$ or $\mu(3)$, a path inside the loop $\{\mu(1), \mu(2), \mu(6), \mu(3)\}$. One is mainly interested in an optimum path leading from μ^0 to the goal(s). This path can be short or long, expensive or inexpensive. It is necessary to introduce the concept of a cost, $\beta(\tau_k)$, for τ_k. If for all edges forming the path the corresponding costs, $\beta(\tau_k)$, are of a unitary value, the overall cost of a walk is proportional to its length, i.e., to the number of intermediate nodes generated.

The cost $\beta(\tau_k)$ is a measure of effort necessary for applying each τ_k, a kind of penalty function weighting the feasibility or probability of each edge. The real cost in dollars of a certain chemical reaction τ_{ab} that transforms a molecule μ_a into another molecule μ_b, or the reaction enthalpy $\Delta H(a,b) = H(a) - H(b)$ are examples of cost.

When a certain parent node gives rise to more than one subgoal, like vertex $\mu(3)$, for example, and there are no unitary cost conditions, then different edges, represented by three different transformations (τ_k, τ_l, τ_m), will be weighted differently and related to their costs, $\beta(\tau_k)$, $\beta(\tau_l)$, and $\beta(\tau_m)$, respectively. In organic chemistry, for example, we could exemplify this abstract discussion by calling the three different transformations τ_k, τ_l, and τ_m the reaction

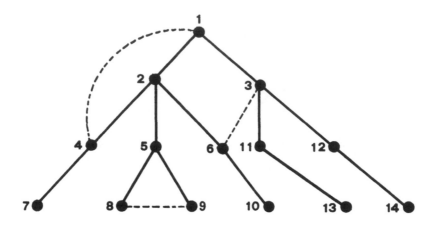

FIGURE 3. Schematic representation of a graph. It differs from a tree in that it contains looping paths between nodes.

mechanisms and the costs $\beta(\tau_k)$, $\beta(\tau_l)$, and $\beta(\tau_m)$ the activation energies. The optimal path is not necessarily the shortest; only an evaluation of the overall path can determine if this is the case. The cost itself is in most cases a complicated multiparametric function containing a formalized description of all intrinsic aspects of τ, but also of μ_n and μ_m, within a path segment $\tau(\mu_m) \rightarrow \mu_m$. The features of the descendant node μ_m must be taken into account, as they can prove dominant when deciding to either generate the node and proceed the walk or not to generate it at all, stop the walk, and change the search direction. To understand this consideration, imagine that a certain chemical reaction, contained formally in τ and showing a known cost $\beta(\tau)$ (including enthalpy, reaction time, product yield, experimental cost, etc.), could be very attractive from a chemical point of view, with $\beta(\tau)$ taking a favorable low value; however, inspection of the product molecule, μ_m, could reveal, for example, that it contains some functional groups which are light sensitive or carcinogenic or shows other features in contrast with commercial and regulative requirements or with technical constraints. Thus, the step $\mu_n \rightarrow \mu_m$ is heavily penalized due to an evaluation of parameters external to τ. The automation of such strategic judgments can sometimes be realized in a computer, but the extent varies from case to case and no general solution can be given.

2. Generating Paths: Breadth-First and Depth-First Searches

There are two principal methods for generating nodes of a tree or graph: the breadth-first search and the depth-first search.[2] The former expands the nodes of a tree in the order in which they were generated. The breadth-first search working principle is illustrated here.

The main steps of the breadth-first search working principle are

First iteration
1. Take initial state $\mu^0 = \mu_{01}$ and put it into vector NEW; NEW $= (\mu_{01})$, OLD $= (\ \ )$.
2. If NEW is empty, quit the program.
3. Take first node μ_{ij} in NEW and place it in vector OLD; NEW $= (\ \ )$, OLD $= (\mu_{01})$.
4. Generate from μ_{ij} all m proximate descendants. Place the descending nodes in vector NEW; NEW $= (\mu_{11}, \mu_{12}, \ldots, \mu_{1m})$, OLD $= (\mu_{01})$.
5. If a descendant is a goal, exit the program; otherwise, go to 2.

Second iteration
2. If NEW is empty, quit the program.

3. Take first node μ_{ij} in NEW and place it in vector OLD; NEW = $(\mu_{11}, \mu_{12}, \ldots, \mu_{1m})$, OLD = (μ_{11}, μ_{01}).
4. Generate from μ_{ij} all n proximate descendants. Place the descending nodes at the end of vector NEW; NEW = $(\mu_{11}, \mu_{12}, \ldots, \mu_{1m}, \mu_{21}, \ldots, \mu_{2n})$.
5. If a goal is achieved, exit the program; otherwise, go to 2 (another iteration).

The described procedure does not involve cost evaluation. If this is required, one must include a cost function in the node generating step.

The special character of the breadth-first method is that the problem space is explored in a broad fashion. From each nth-generation node all adjacent nodes are generated first; then, following the sequence of their birth, all proximate $(n + 1)$th generation nodes are produced; and so on. In the tree of Figure 2 this chronology is demonstrated by the creation of the states $\mu(2)$ and $\mu(3)$ from $\mu(1)$, then of $\mu(4)$, $\mu(5)$, $\mu(6)$, and of $\mu(7)$, $\mu(8)$; then $\mu(9)$ is generated, and $\mu(10)$, $\mu(11)$; then comes $\mu(12)$, as well as $\mu(13)$, and finally $\mu(14)$.

The depth-first method employs an opposite strategy, searching deeply into the problem space first, the action of fanning out in breadth being only a secondary consequence of this exploring mechanism.

The descendants of a node are placed at the beginning of vector NEW, causing the first node of a new generation to be expanded first. At step 4 of the above procedure, vector NEW will contain the following sequence of elements:

$$NEW = (\mu_{21}, \mu_{22}, \ldots, \mu_{2n}, \mu_{11}, \ldots, \mu_{1m})$$

The boundary number of generations that must be reached before stopping the depth-first search is given by a predetermined number, the depth bound. The search then proceeds from another equivalent node in a momentary in-breadth mode. The chronology of a depth-first walk through Ω is again exemplified in Figure 3. The starting node leads to $\mu(2)$ and $\mu(3)$, the former generating $\mu(4)$, $\mu(5)$, and $\mu(6)$; $\mu(4)$ is the starting point of a new generation ending in $\mu(7)$. Similarly, $\mu(5)$ leads to $\mu(8)$ and $\mu(9)$, and $\mu(6)$ gives rise to $\mu(10)$. In our example we set 3 as the generation depth boundary; only at this level can $\mu(3)$ be recalled and processed to expand the other branch of the tree down to $\mu(14)$.

When dealing with graphs one must be aware that a node can be generated in more than one manner (multiple paths): marking generated nodes in a unique way avoids redundant generation. (This is very important in the coding and identification of molecular structures, as demonstrated later.)

An example will now illustrate the necessity of introducing some heuristic rules into a goal-searching procedure. So far we have discussed two major approaches in their purely mechanistic schemes. In real situations, additional strategies must be added to make the search effective and practicable. Otherwise, the described methods appear just as two variations of a random-search algorithm. (Sooner or later they lead to a success, but devoid of any better path selection capability they may get stuck in a hopeless combinatorial explosion of paths and states.)

Let us take four amino acids, A, B, C, and D, which are joined to form a hypothetical tetrapeptide, DCAB. Suppose that a researcher who is interested in mutagenicity and is studying phylogenetic trees of proteins wants to know how many mutagenic paths lead to another peptide, ABCD, that seems to show activity similar to that of DCAB. He is also interested in finding the shortest paths.

The computer has to expand the initial state, DCAB, according to some transformation rule and, going from generation to generation of the descending peptides, it must stop whenever an alignment ABCD (the goal) is realized. The rule is to apply a permutation operator π such that $\pi(A, B) = (B, A)$.

Starting from DCAB (see Figure 4) and exploiting the six possible permutations, six new amino acid sequences are obtained in the first generation. This situation is the same for both the breadth-first and the depth-first searches. The step to a second generation applying the breadth-first mechanism gives rise to six tetrapeptides for each parent subgoal: in total, 36 second-generation tetrapeptides are formed. With a depth-first search, the sequence DCAB would be the result of the permutation $\pi(C,D)$ in the sequence CDAB. However, this is the sequence of the initial state, and an appropriate check prevents the definitive creation of this redundant second-generation node. The first of five valid subgoals derived from CDAB is ADCB. A third generation becomes necessary to obtain a goal state. According to the depth-first search algorithm, the first component of the second-generation vector is processed first; that is ADCB. Permutation of D and B results in ABCD.

Figure 4 shows that (1) many redundant sequences are generated which must be eliminated by a special routine in an eventual computer program and (2) whatever method is chosen, the solution is obtainable only after three generation steps.

It must be noted that from all branches in the graphs there are two or more paths leading to the target sequence.

With the breadth-first technique, $6 + 36 + 1 = 43$ transformations were carried out before a goal state was encountered at the end of a walk starting at DCAB, going over ADCB to ABCD. There is no shorter path. Using a depth-first technique, only $6 + 6 + 1 = 13$ transformations were necessary. The impression could arise that roughly a third of the computing time is demanded for the simulation with the latter method. This suspicion is erroneous, as we are still extending the search in a random manner and a goal state could be reached equally well in a breadth-first manner in some other case. In more complex problems, for example, the walk can protract inconveniently in a depth-first search into distant meanderings through the problem space, whereas an early, wider perception of the initial state's neighborhood may be far more useful. In addition, as in the above example, if several possible paths leading to the same goal exist, a breadth-first search offers a strategically better panoramic view of relevant nodes. This calls for some additional evaluation strategy.

It could be nonsense to generate all isomers from a given molecular formula in order to correlate one structure with some spectral data at a later time. It is much better, in a heuristic approach, to generate only those structures containing *a priori* the substructures and functional groups for which clear evidence has been found of their being part of the unknown structure. In this way one reduces the number of irrelevant nodes considerably. Heuristic programming consists of an apposition of (mostly empirical) adjunct information rendering the exploratory computer simulation "intelligent".

If one is not interested in the complete set of concurrent paths generating the whole graph, but just in finding one successful path, it is easy to individuate which information would optimize the search in our elementary tetrapeptide example. In the first generation we note that four of the six sequences have one amino acid already placed in a position matching the goal's pattern, while all amino acids of the two remnants are in incorrect positions. Each of the latter further generates six new alignments, of which only two become goal states after one additional permutation. However, the previous four sequences containing one correctly placed amino acid, after expansion into the second level, show three new sequences with a configuration distance of just one permutation from the goal configuration. All of the mentioned sequences have two correctly placed elements. Thus, a heuristic rule would be to take into account at every generation level the momentary goodness of fit between subgoal and goal. In our case, only the first-generation sequences with one element already matching the goal pattern should be processed, neglecting the other two. This leads into a region of Ω which has a higher probabilistic density of states with increased goodness of fit. Another rule could be to forbid application of π on a pair of elements containing one

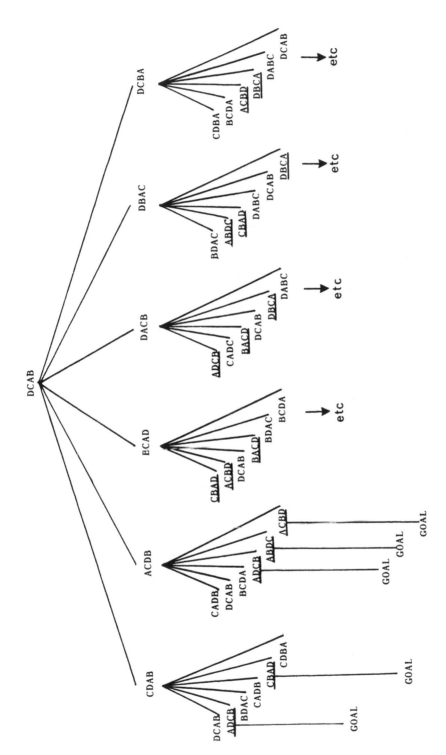

FIGURE 4. A simple example of goal identification. A tetrapeptide, DCAB, is taken as the initial state for the search of the minimal number of steps necessary to obtain a goal tetrapeptide, ABCD.

element that already is positioned correctly. For example, when processing the sequence ACDB, the first three nodes generated are CADB, DCAB, and BCDA, in which A is displaced from its former correct position. Avoiding their creation means applying π only on the subsequence CDB of ACDB, an operation that results in the straightforward generation of ADCB, an immediate precursor of a goal. This focused processing would not be possible for CDAB because no element is placed correctly yet, and the permutation operator must act on the complete array of all four amino acids (CDAB) instead of only three, causing twice as many nodes to be expanded (six instead of three). We see that two simple rules of a heuristic nature (obviously more can be added) provide for a drastic optimization of the search performance.

The goodness of fit between a certain subgoal and the goal was given here simply by the distance of a node from its final goal. Here we dealt with a topological distance for which a specific metric exists and which plays an important role in the formal treatment of chemical reactions and in the statistical analysis of chemical data. The minimum distance between DCAB and ABCD is 3, between DBCA and ABCD is 1, and between ABCD and ABCD is 0 within a permutation metric. Attributing to each symbol A, B, C, and D a specific number, euclidean distances can be calculated for the various pairs of tetrapeptides.

It is one of the fundamental tasks of a researcher developing AI simulation programs in chemistry, as well as in other sciences, to tune his procedures for problem space exploration by including heuristic solving methodologies whenever necessary.[10]

REFERENCES

1. **Stoll, R. R.,** *Set Theory and Logic,* Dover Publications, New York, 1963.
2. **Margaris, A.,** *First Order Mathematical Logic,* Blaisdell Publishing, 1967.
3. **Curry, H. B.,** *Foundations of Mathematical Logic,* Dover Publications, New York, 1977.
4. **Slagle, J. R.,** *Artificial Intelligence: The Heuristic Programming Approach,* McGraw-Hill, New York, 1971.
5. **Nilsson, N.,** *Problem Solving Methods in Artificial Intelligence,* McGraw-Hill, New York, 1971.
6. **Kuester, J. and Mize, H. J. H.** *Optimization Techniques with Fortran,* McGraw-Hill, New York, 1973.
7. **Massart, D. L., Dijkstra, A., and Kaufman, L.,** *Evaluation and Optimization of Laboratory Methods and Analytical Procedures,* Elsevier, Amsterdam, 1978.
8. **Lindsay, R. K., Buchanan, B. G., Feigenbaum, E. A., and Lederberg, J.,** *Applications of Artificial Intelligence for Organic Chemistry,* McGraw-Hill, New York, 1980.
9. **Harary, F.,** *Graph Theory,* Addison-Wesley, Reading, MA, 1972.
10. **Pierce, T. H. and Hohne, B. A., Eds.,** *Artificial Intelligence Applications in Chemistry,* ACS Symp. Ser., Vol. 306, American Chemical Society, Washington, D.C., 1986.

Chapter 4

MOLECULAR MODELING

I. FUNDAMENTALS OF MOLECULAR MODELING

A. INTRODUCTION

Molecular modeling is a general, all-comprising term including perception, manipulation, physicochemical parameterization, and visual reproduction of molecular structures by a computer.

Historically, the first impact among traditional chemists was certainly found in two areas: in the graphical representation of chemical structures obtained from crystallographic data and in the study of the conformations of flexible molecules using molecular mechanics techniques.

The principal information about molecular structure can be perceived and understood more easily when presented in the chemist's most familiar language: by a picture of its structure by means of molecular graphics techniques, rather than by a list of numbers. The human brain is the best pattern recognizer, and a picture is better than a thousand words.

The availability of increasingly more powerful computers and the parallel development of computer graphics have elected molecular modeling one of the driving forces in computer chemistry research. It is a substantial, probably unsubvertible tool in advanced investigations in the expanding fields of drug design[1-8] and, in the near future, of materials science. Its undeniable success, especially in the chemical and pharmaceutical industries, is rooted mainly in providing user-friendly interaction even for the non-computer-oriented chemist, for whom an easy access has been paved to the use of this methodology, casting a new perspective on research strategy. The popularity of molecular modeling programs has increased greatly in the last few years due to the creation of software conformed to the size of mini- and microcomputers, which makes molecular modeling accessible to a large number of even small laboratories.

Such programs and their required corollary hardware must be regarded as additional instruments in a laboratory outfit. This is the only way to justify them; exaggerated enthusiasm as well as blind rejection are equally misplaced.

With computer graphics one can draw, visualize, and freely manipulate three-dimensional (3-D) structures, an operation relegated in the past to the slow and ineffective handling of clumsy plastic models. Computer-generated, virtual 3-D models open the way to opportunities for structural manipulation almost impossible with real models, like molecular shape comparison and molecular docking, emphasizing that the times when drug research was run on a "hit-or-miss" principle are slowly fading away beyond the scientific horizon.

But visualization is certainly not the only achievement of molecular modeling: designing a new drug requires, after its 3-D structural generation, the determination of one or more possible conformation energy minima. Adding other quantities like atomic charges, molecular volume, and area, as well as other physicochemical descriptors, relationships can be established between the structure and properties of drugs — and in more difficult cases, between drugs and their hypothetical biological receptors.

For this reason, a great deal of effort has been invested in finding fast empirical models for the computation of many molecular descriptors used in drug design and, more specifically, in quantitative structure-activity relationships (QSAR).

Before dealing in depth with the mentioned themes, we must not forget that, when talking about AI applied to chemistry, the problem states perceived by the computer are molecular structures which are "condensed" in symbols, each symbol encoding its own quantum of chemical information. It follows that molecular modeling must be primarily a kind of communication language between man and computer, the imperative interface between human-comprehensible symbolism and the mathematical description of an inherently invisible entity, the molecule. The very heart of the problem in the early days of this endeavor was to make a computer perceive and recognize a molecular structure out of its symbolic representation. Conversely, the computer must be able to return chemical answers in an equivalent chemical symbolism.

B. GENERATION AND REPRESENTATION OF TWO-DIMENSIONAL MOLECULAR MODELS

Many methods have been developed to introduce information about a two-dimensional (2-D) or 3-D molecular structure into a computer. They can be classified into two categories: the first category, historically the oldest, uses more or less sophisticated linear sequences of typed-in programmer-defined symbols to reproduce the topology, i.e. the connectivity relations of a 2-D molecular model. A 3-D molecular model can then be generated by processing the 2-D model with empirical force-field calculations.

The second category involves graphical input methods. The molecule, or better yet the global molecular symbol, is generated directly on a graphics terminal by specific generating and assembling commands. There is no need for any precoding of atoms or bonds. Methods are available for 2- and/or 3-D model generation.

The input symbols, graphical or alphanumerical, are translated inside the computer into an internal representation by programmed rules forming the core of the coding/decoding procedures, rules that simulate human perception of a molecular image. This representation is a mathematical codification used internally by the program for the actual structural manipulations; it is normally not visible to the user. From the internal representation, the same rules are used to create an equivalent output symbolism, such as a graphic display of chemical structures.

The following scheme summarizes the communication mechanism between chemist and computer:

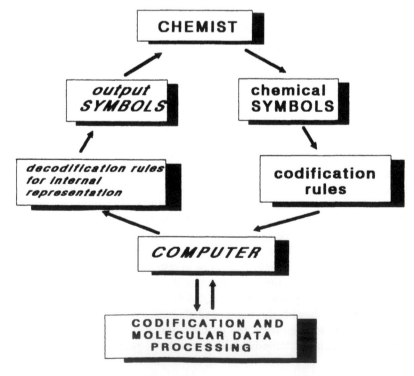

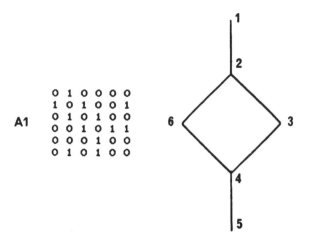

$$
\mathbf{A1} \quad
\begin{matrix}
0 & 1 & 0 & 0 & 0 & 0 \\
1 & 0 & 1 & 0 & 0 & 1 \\
0 & 1 & 0 & 1 & 0 & 0 \\
0 & 0 & 1 & 0 & 1 & 1 \\
0 & 0 & 0 & 1 & 0 & 0 \\
0 & 1 & 0 & 1 & 0 & 0
\end{matrix}
$$

FIGURE 1.   The adjacency matrix corresponding to the molecular graph in one particular numbering scheme.

The internal representation is generally more complex than the symbolic input form, as it contains the information about bonds and atoms in explicit form; the internal representation is suitable for logical and algebraic processing (a major difference in the strict sense from chemical symbolism, which is not directly mathematically tractable).

1. Topological Encoding

The topology of any chemical structure, whenever regarded as a graph, can be presented in a formal manner by an adjacency matrix.[9] Their matrix elements are defined by the relations

$$
a_{ij} = \begin{matrix} 1 & \text{for adjacent vertices} \\ 0 & \text{otherwise} \end{matrix}
$$

This is the simplest symbolism for communicating a 2-D molecular model to a machine, but at the same time it is already one possible mathematical representation of the structure. The molecular structure shown in Figure 1 has an associated adjacency matrix, A_1.

However, the actual form of an adjacency matrix depends on the numbering of the atoms; if the molecular graph in Figure 1 is renumbered arbitrarily, a second adjacency matrix, A_2, is equally valid (Figure 2). Different adjacency matrices can therefore originate from the same graph, and recognizing that two different-looking topological matrices are coding the same molecular graph is not easy. There are in fact $n!$ different numerations (neglecting reduction due to symmetry) of a graph with n vertices, associated with corresponding adjacency matrices. To check for equivalence of two graphs it is necessary to permute rows and columns of the matrices, which is feasible when the number of vertices is small, but becomes a lengthy procedure for large molecular graphs, even with a computer.

The characteristic polynomial (CP) of an adjacency matrix is defined as the expansion of the determinant of the matrix $A - \alpha I$,

$$
CP = \det (A - \alpha I)
$$

where I is the identity matrix and α the eigenvalues. For example, propane has the following characteristic polynomial (C-skeleton only):

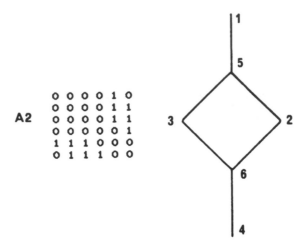

$$
\mathbf{A2} \quad
\begin{array}{cccccc}
0 & 0 & 0 & 0 & 1 & 0 \\
0 & 0 & 0 & 0 & 1 & 1 \\
0 & 0 & 0 & 0 & 1 & 1 \\
0 & 0 & 0 & 0 & 0 & 1 \\
1 & 1 & 1 & 0 & 0 & 0 \\
0 & 1 & 1 & 1 & 0 & 0
\end{array}
$$

FIGURE 2.   Renumbering of the structure showing in Figure 1 leads to a different adjacency matrix.

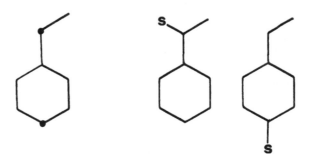

FIGURE 3.   The graphs shown, which differ in substitution at isospectral nodes, yield the same eigenvalues for their characteristic polynomial. They are isospectral graphs.

$$
\begin{array}{ccc}
1 & 2 & 3
\end{array}
\;\cdot\!\!-\!\!\cdot\!\!-\!\!\cdot\; \rightarrow \mathbf{A} =
\begin{pmatrix}
0 & 1 & 0 \\
1 & 0 & 1 \\
0 & 1 & 0
\end{pmatrix}
\rightarrow (\mathbf{A} - \alpha\mathbf{I}) =
\begin{pmatrix}
-\alpha & 1 & 0 \\
1 & -\alpha & 1 \\
0 & 1 & -\alpha
\end{pmatrix}
$$

It follows for CP that

$$
-\alpha(\alpha^2 - 1) - 1(-\alpha - 0) - 0 = 0
$$

the eigenvalues of this polynomial being $\alpha_1 = 0$ and $\alpha_{2,3} = \pm 2$. The eigenvalue array is called the spectrum of eigenvalues. It is known that the CP of the adjacency matrix of a given graph is unfortunately *not* uniquely related to a single topological structure. There are nonisomorphic graphs with identical CPs, consequently having the same spectrum of eigenvalues.[10,11] The existence of such isospectral, nonisomorphic graphs is a major shortcoming in the use of the CP eigenvalue as a compact, topological molecular name. For example, two isospectral molecular graphs are given in Figure 3 with two styrene-like π-system graphs. They have been generated by substituting any graph S at so-called isospectral points, denoted by bold vertices.[12]

An adjacency matrix is a mathematical construction expressing the presence of edges

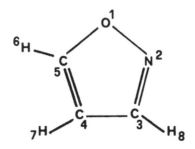

FIGURE 4. Isoxazole is numbered arbitrarily to illustrate the generation of a *BM* matrix.

between vertices on a molecular graph. An endemic limit to this description is its monochromic character: the vertices are all equal and do not reproduce the polychromic variety of atoms in a real molecule. Also, no information about the bond order between atoms is given, as all edges in the graph are equivalent. An improvement comes from using an atom connectivity matrix, which is closer to a chemical description than the adjacency matrix used by mathematicians in graph theory. The atom connectivity matrix consists of the bond order in the off-diagonal elements and of the atomic number in the diagonal elements. Hydrogen cyanide, $H-C\equiv N$, has the following atom connectivity matrix (dependent on the numeration scheme):

	H	C	N
H	1	1	0
C	1	6	3
N	0	3	5

Both the adjacency matrix and the atom connectivity matrix are square, symmetrical matrices of n^2 elements. They are redundant in their informational content, not satisfying the request for an optimized core memory occupation and therefore not well suited for an internal representation. They are certainly too unwieldy to be typed in from a keyboard (50 atoms in a molecule would mean 2500 entries). The question of how much space a program requires is not outdated, as some people might think when invoking the growing capacity of computers. It is true that memory size grows constantly, but it is equally true that the complexity of our chemical problems grows, too. The pioneer years of molecular modeling offered the display of short-chain alkanes on screen terminals, and that was a great achievement. Today we want to display the dynamics of intercalating mutagens in DNA helices, or protein-substrate docking phenomena. When programming such kinds of software an eye must always be kept on space requirements, and that means working with an optimized data structure.

Considering that a molecule has approximately as many bonds as atoms, a compact way to describe the polychromic topology of a molecule is given by a condensed bond-electron matrix (*BM*). The diagonal of the former connectivity matrix is contained in a separate vector, and the information about bond orders and connections is compressed into a $2m \times 3$ matrix (m being the number of bonds). Each column represents a bond, the first two elements being the numbers of the bond partners. The third entry is the bond order. As an example, the *BM* matrix for isoxazol is shown here (see Figure 4).

Atom numbering   1 2 3 4 5 6 7 8

Diagonal   8 7 6 6 6 1 1 1

pointers

↙ ↘

$$BM \text{ matrix} \quad = \quad \begin{pmatrix} 1\,1\,2\,2\,3\,3\,3\,4\,4\,4\,5\,5\,5\,6\,7\,8 \\ 2\,5\,1\,3\,2\,4\,8\,3\,5\,7\,1\,4\,6\,5\,4\,4 \\ 1\,1\,1\,2\,2\,1\,1\,1\,2\,1\,1\,2\,1\,1\,1\,1 \end{pmatrix}$$

The careful reader also will have noticed that a *BM* notation contains a twofold redundancy: each bond appears twice. This, however, is a price than can be paid considering that (1) the perception algorithm takes a much simpler form and (2) the characteristic feature of molecular graphs, their nondirected edges, is preserved. A bond *K–L* is and must be equivalent to *L–K*. The first point is demonstrated easily when the computer has to perceive the complete neighbor sphere of a certain atom *i*. Pointers at the start $[p_s(i)]$ and at the end $[p_E(i)]$ of columns with the same atom *i* in the first row permit a faster and more direct access to all its $p_E(i) - p_s(i) + 1$ neighbors (located in the second row of a *BM* matrix).

This is a very useful internal representation, although incomplete in the form just presented (free electron pairs, ring atoms, and formal charges must be added). The number of matrix elements in a *BM* code only increases linearly with the number of atoms, in contrast to the quadratic increase in a connectivity matrix notation.

A chemist using a molecular modeling interface program wants an even shorter code to input his structures. A complete description therefore must be replaced by some quick symbolism, which requires an interpreter inside the interface to decode the user-input symbol string.

A number of alphanumeric input methods have been published, among them the Wisswesser line notation,[13,14] which became very popular before the birth of more user-friendly direct graphic input methods.

The program system for the simulation of organic reactions (EROS) uses, for example, a linear alphanumeric input mode, the MOLIN code, as well as a graphical input facility.[15] Taking molecule A as an example and arbitrarily numbering its atoms, the following input string is obtained:

The MOLIN code is as follows: HF 1 S 2 O 3 C 9 J * (1 5 D 6 7 D 8 1) (5 4 3) 4 D 2 8 9 * *. The HF symbol alerts the program that all hydrogen atoms not explicitly given by the user in the input must be attached autmically to the free valences of the heteroatoms and filled into the bond relations of the *BM* matrix. The number of the first heteroatom of a specific kind is then added. For carbon, as an example, it suffices to enter "3 C" (and not "3 C 4 C 5 C . . . ") until the next type of heteroatom, atom #9 (iodine), is encountered.

An * is the symbolic mark for ending the input of an atom type. The nonredundant coding of bonds is simple: inside of parentheses are listed chains of connected heteroatoms, a ''D'' standing for a double bond. This is a purely symbol-oriented convention of linear input which can be changed at any time. It is immediately translated into a full internal data structure suitable for mathematical manipulation. In principle, the internal notation could serve to reproduce an analogous output symbolism, but this would be quite cumbersome to read. In all cases of computer chemistry programs processing molecular structures the output symbolism consists of displays of their structures.

Another example of 2-D structural codification is given by the CONOL-II convention found in the synthesis design programs SCANSYNTH and SCANPHARM.[16,17] It is illustrated here on acetoxy derivative B:

The CONOL-II code is 1C1–C2–2C1–C2–C2–1C1,1–C3,2–O–3C–C3,3=O. 1C1 stands for a carbon atom bonded to one hydrogen an another substituent; C2 means a methylenic group, CH_2; 2C1 indicates that a second CH group exists requiring an additional substituent; and the terminal symbol 1C1, equal to the first, defines the ring closure atom. The substituents are coded as follows: a methyl group (C3) in position 1, a group $O–C–CH_3$ for atom ''2C1'' in position 3, and, finally, a double-bonded oxygen atom for the free valences of the acetoxy carbon atom. This stenographical description is decoded by the computer and rearranged into an internal matrix representation. Within the program a relationship exists between each codified substructure and a substructural index number. These indices form a so-called functone connectivity matrix (FCM), a functone being a primitive substructure contained in a data base of predefined substructures. For molecule B an FCM (i.e., a square symmetric matrix that in principle could be transformed into a more compact BM-like form) can be written as follows:

$$
FCM(\mathbf{B}) =
\begin{array}{c|ccccc}
 & 1 & 16 & 16 & 18 & 23 \\
\hline
1 & 0 & 101 & 0 & 0 & 102 \\
16 & & 0 & 0 & 0 & 0 \\
16 & & & 0 & 1 & 0 \\
18 & & & & 0 & 1 \\
23 & & & & & 0 \\
\end{array}
$$

Entries 101 and 102 mean the first and second substituents, 16 and 23, on substructure 1. Similarly, from the input string other vectors containing information about ring atoms, size, heteroatoms, etc. are obtained. This coding method differs conceptually from the MOLIN code because it is designed to operate with a data base of preexisting substructures, whereas the EROS input philosophy is devoid of any previous cognition of assembled atoms. The EROS input is specifically a bond-oriented input due to its prefixed goal: it is the input of a synthesis design program and must provide for the labeling of specific breakable bonds.

A first brilliant achievement in computer chemistry programming was given by the interactive input method of the structure elucidation programs of the DENDRAL project (see Chapter VI). In DENDRAL the definition of rings, chains, double bonds, and all kinds of substructures is done semantically with symbolic words. For example, to assemble the topological structure of cyclobutanone, the following session has been recorded with the CONGEN program (capital letters are computer answers while lower case ones are human input, and the session print is self-explanatory):

session record:

#define substructure cbone
(NEW SUBSTRUCTURE)
>ring 4
>branch 1 1
>join 1 5
>draw atnumbered
SUBSTRUCTURE CBONE:

>atname 5 o
>hrange 2 0 2 3 0 2 4 0 2
>show
SUBSTRUCTURE CBONE:

ATOM#	TYPE	NEIGHBORS	HRANGE
1	C	5 5 4 2	0—2
2	C	3 1	0—2
3	C	4 2	0—2
4	C	1 3	0—2
5	O	1 1	

>done
CBONE DEFINED

As will be shown later, this is an input mode tailored especially for computer-assisted generation of unknown structures from substructural fragments, as clearly evident from the range of hydrogen atoms bonded to atoms 2, 3, and 4: from zero to two hydrogen atoms can be bonded, thereby leaving possible open sites of attachment for other intervening substructures.

2. Ring Perception

Perception of rings in molecules is a major challenge for any AI approach in chemistry. Based on the various internal representations typical for each system, special algorithms have been developed for the computer-inferred determination of chemically relevant rings in a molecule described only by its topology. Chemically relevant means that ring size can be correlated to spectral or synthetic features. Thus, real rings can be separated from virtual rings, which have only topological existence, but are not an experimentable reality. A correct perception of these relevant rings is fundamental for the performance of synthesis design programs where strategic bonds in rings must be formed or broken in elucidation of chemical structures, in automated interpretation of spectra, and in the retrieval of structures or substructures in on-line data bank research sessions.

One becomes aware of the complexity of the problem when determination of the exact number of rings (real and virtual) in cubane is desired:

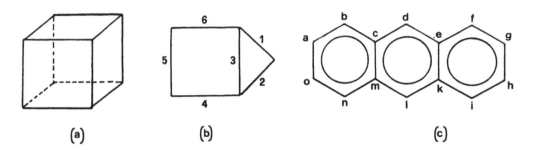

The backbone of a ring perception algorithm will now be discussed, the aim of which is to find the complete set of topologically existent rings.

Since computers with optical pattern recognition capabilities are not readily available today, although progress has been made toward creating a kind of visual intuition by high-speed image recognition in many fields, the perception of a ring must follow a stepwise procedure here. The number of rings in a graph is equal to or larger than the Frerejaque number, given by Equation 1:[18]

$$\text{number of rings} = \text{number of bonds} - \text{number of atoms} + 1 \tag{1}$$

For anthracene, one obtains $16 - 14 + 1 = 3$; for cubane, $12 - 8 + 1 = 5$; and for bicyclo(2.1.0) pentane, $6 - 5 + 1 = 2$.

For the latter compound, using the bond numbering shown above, two independent vectors ($R1$ and $R2$) are given in binary form by

$$R1 = (1,1,1,0,0,0); \quad R2 = (0,0,1,1,1,1)$$

where we mark a bond member of a ring system with a logical "1" (*true*) and the opposite with a logical "0" (*false*). $R1$ and $R2$ are the basis vectors of the fundamental set of the smallest rings of the bicyclic compound. They point at two problem states, R1 and R2, in the problem subspace containing all possible rings obtainable with six bonds.

With the logical operation of Equation 2,

$$R3 = R1 \text{ XOR } R2 = (1,1,0,1,1,1) \tag{2}$$

one obtains a third, linearly dependent vector ($R3$) for the five-membered, virtual ring, one not predicted in the Frerejaque number.

A definition of a special fundamental set of rings, the smallest set of smallest rings (*SSR*),[19,20] has been introduced. All rings are thereby ordered according to size. The smallest ring is attributed to the *SSR*; successive rings are included if they are linearly independent from the previously selected ones. The *SSR* algorithm delivers a number of fundamental rings equal to the Frerejaque number. A bridge from this formal digression to real chemistry

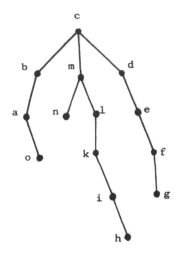

FIGURE 5. The spanning tree of nodes
generated during the search for the smallest
set of smallest rings in structure c.

is thrown, for example, by the dependence of the ring strain energy on the size of the
smallest rings, which plays a role in experimental organic synthesis. The envelope ring *R3*
certainly will not determine the ring strain energy of the molecule. It is a heuristic definition
to call the rings in an *SSSR* a set of fundamental rings for organic chemistry. Different
fundamental sets have been proposed.[21-25]

The determination of the various atom sequences R for all rings evolves over a spanning
tree of search steps for closing paths in the molecular graph. We define the path length, or
distance $D(i,j)$ between two atoms, i and j, as the number of bonds necessary to connect i
to j. A path D^0 starting from a given atom i and returning back to i without passing any
bond twice must be found such that

$$D(i,i) = D^0 = \text{minimum}$$

This statement forces a ring to have at least three edges: a walk from a certain vertex i to
any adjacent one j and back to i is a walk with $D(i,i) = 2$, but edge (i,j) is passed twice,
disobeying the above proposition. You shall be introduced to another formal definition of
rings consisting of two atoms after the concept of unsaturation has been discussed.

We define as the Dth proximity sphere of an atom i the set $\{j_{d1}, j_{d2}, \ldots, j_{dk}\}$ of all k
atoms having the same distance, $|D_d| = D(i,j_{dn})$, from i ($n = 1,k$).

The anthracene skeleton with the atoms labeled as shown above will serve to demonstrate
the principles of the algorithm.[26] The spanning tree starts from an arbitrary initial vertex
(atom) of the graph, e.g., atom c. A breadth-first search provides for the generation of
subsequent Dth-level proximity atom sets (see Figure 5). (In a *BM* topological description
this is performed easily, as all neighbors of a certain atom are aligned compactly as pointed
second-row matrix elements.) In the first generation, with $D_1 = 1$, the three paths going
to atoms b, m, and d are spanned. They form the subset $\{b,m,d\}$ of atoms of the first
proximity sphere. Each of them is the father of a second-generation set of vertices ($D_2 =$
2), consisting of the following four atoms: a, n, l, and e. In the next step, the first node to
be processed following the breadth-first search method is node a. It generates node o, which
is also generated if the next parent node n is processed. Since o has been created already
and any redundance must be avoided, a termination criterion is reached to stop the left-most
path at o and the left branch of the middle path at n. The edge (o,n) is called the ring closure

bond (*RCB*). Proceeding accordingly, the complete tree is spanned down five levels, ending with atom *h*. Two other *RCB*s occur, between atoms *k* and *e* and between *h* and *g*. All *RCB*s are ordered after increasing *D* from origin *c*. At every *RCB*, starting from the farthest, a walk back to *c* is initiated to determine the *SSSR*. As soon as a common node or a node belonging to the most proximate *RCB* is found, a "smallest" ring and its atoms are revealed and the procedure repeated anew. The following real, chemically relevant rings are detected:

R1: (a,b,c,m,n,o); R2: (c,d,e,k,l,m); R3: (e,f,g,h,i,k)

As the detection of any new ring involves the use of different *RCB*s, the dimension of the ring vector space is always augmented by at least one, and the rings of the *SSSR* are all independent vectors of this space. Now it is possible to combine these basis vectors to construct all possible rings of anthracene. Using a binary notation we can write

$$\begin{array}{rl} & a\ b\ c\ d\ e\ f\ g\ h\ i\ k\ l\ m\ n\ o \\ R1 = & (1\ 1\ 1\ 0\ 0\ 0\ 0\ 0\ 0\ 0\ 0\ 1\ 1\ 1) \\ R2 = & (0\ 0\ 1\ 1\ 1\ 0\ 0\ 0\ 0\ 1\ 1\ 1\ 0\ 0) \\ R3 = & (0\ 0\ 0\ 0\ 1\ 1\ 1\ 1\ 1\ 1\ 0\ 0\ 0\ 0) \end{array}$$

With the Boolean operator **OR** the following equations hold:

$$\begin{array}{rl} R4 = R1\ \mathbf{OR}\ R2 = & (1\ 1\ 1\ 1\ 1\ 0\ 0\ 0\ 0\ 1\ 1\ 1\ 1\ 1) \\ R5 = R2\ \mathbf{OR}\ R3 = & (0\ 0\ 1\ 1\ 1\ 1\ 1\ 1\ 1\ 1\ 1\ 1\ 0\ 0) \\ R6 = R1\ \mathbf{OR}\ R2\ \mathbf{OR}\ R3 = & (1\ 1\ 1\ 1\ 1\ 1\ 1\ 1\ 1\ 1\ 1\ 1\ 1\ 1) \end{array} \quad (3)$$

The envelope ring *R*6 is an example of the family of non-*SSSR* members which, although not having synthetic importance (one does not at once synthesize this 14-membered ring, but starts from preexistent smaller cyclic synthetic units), nevertheless have a realistic importance in spectroscopy: the size of the envelope rings of alternating aromatic systems correlates very well with the promotion energy $E\pi(\mathrm{b} \rightarrow \mathrm{a})$ of the electronic spectrum, visible in the ^1L_a absorption bands. The envelope is in this case a measure of the extension of the delocalized π-electron system. The vectors with the *SSSR* ring information are added to the internal representation.

The description of molecular topology given so far has an ambiguity: everything still depends on the initial numbering scheme of the molecular skeleton. When generating structures, a redundancy can be eliminated only if a unique name exists for each graph. Thus, a method should be found to label atoms with one, and only one, possible numbering scheme such that the *BM* matrix description and the ring vectors will be specific for each structure. We can consider this to be a mathematical name. Such names are called canonical names. Without a canonical numbering scheme, different arbitrary numerations lead to different *BM* matrices; for example,

$$\begin{array}{c} 1\ \ 2\ \ 3 \\ \mathrm{H{-}C{\equiv}N} \end{array} \qquad\qquad \begin{array}{c} 1\ \ 3\ \ 2 \\ \mathrm{H{-}C{\equiv}N} \end{array}$$

$$BM = \begin{array}{c} 1\ 2\ 2\ 3 \\ 2\ 1\ 3\ 2 \\ 1\ 1\ 3\ 3 \end{array} \qquad\qquad BM = \begin{array}{c} 1\ 2\ 3\ 3 \\ 3\ 3\ 1\ 2 \\ 1\ 3\ 1\ 3 \end{array}$$

In chemical documentation as well as in structure elucidation and synthesis design programs, such canonical names are used to avoid duplicated creation, storage, or retrieval of structures.

3. Canonical Numbering

Canonical numbering algorithms are heuristic procedures applying human-defined conventions to the discrimination of atoms in different topological environments.

The designer of canonical numeration rules is free to invent whatever tactical moves and tricky descriptions he believes can enhance the speed, the selectivity, and the generality of his procedure. Starting from the simplest description of a graph, the adjacency matrix, some authors have proposed the following solution: if the lines of the binary matrix are rearranged along a single line, the resulting binary string can be understood as the binary code of an integer number.[9] For example,

$$
\begin{matrix}
1 \; 2 \; 3 \\
\text{H–C≡N}
\end{matrix}
\qquad
\begin{pmatrix}
0 & 1 & 0 \\
1 & 0 & 1 \\
0 & 1 & 0
\end{pmatrix} \text{adjacency matrix}
$$

$$
\begin{matrix}
\text{binary number} \\
0 \quad 1 \quad 0 \quad 1 \quad 0 \quad 1 \quad 0 \quad 1 \quad 0
\end{matrix}
\quad \equiv \quad
\begin{matrix}
\text{decimal number} \\
2^7 + 2^5 + 2^3 + 2^1 = 170
\end{matrix}
$$

There is a unique permutation of the atom numbers, resulting in an exchange of rows and columns in the adjacency matrix, that corresponds to the lowest decimal number. Permuting columns 2 and 3, which is equivalent to renumerating hydrogen cyanide as

$$
\begin{matrix}
1 \; 3 \; 2 \\
\text{H–C≡N}
\end{matrix}
$$

leads to the string $0\ 0\ 1\ 0\ 1\ 0\ 1\ 0\ 1 = 2^6 + 2^4 + 2^2 + 2^0 = 85$. This is the smallest number extractable from the HCN adjacency matrix. This procedure, although based on one simple rule, is sometimes too time-consuming for complicated cases, as it appears that the simple permutations $\pi(a,b) \rightarrow (b,a)$ may not suffice to find the absolute minimum. It has been shown that sequences of permutations are required to overcome this deficiency of the simpler procedure.[11] Furthermore, it is very possible that nearly all n! different matrices will be generated before the best one is traced.[27]

Besides this mathematical solution attempt, there are a number of methods based on equivalence classes. Atoms are thereby subdivided into specific classes according to predefined atom descriptors. After each iteration the classes are partitioned further into smaller classes, and in an optimal situation, neglecting constitutional equivalences, the final classes each contain only one atom. The atom descriptors are a quantitative empirical result of the application of user-defined rules to the molecular topology under study. Here the main effort lies in localizing good rules with high discriminating power. The following explains the conceptual lines of how a numeration procedure could be constructed:

- First rule: atoms with higher atomic number Z are numbered first
- Second rule: atoms with fewer hydrogen bond partners are numbered first

Using these rules, the canonical numbering of propionitrile will be

$$
\begin{matrix}
1 \; 2 \; 3 \quad 4 \\
(A) \quad \text{N≡C–CH}_2\text{–CH}_3
\end{matrix}
$$

Generating from this initial state two other states by isomerization (e.g., NH_2–CH=C=CH_2 and CH≡C–CH_2–NH_2), one gets the canonical numberings

$$
\begin{array}{ccc}
 & 1 \quad 3 \quad 2 \; 4 & \quad\quad 3 \; 2 \; 4 \quad 1 \\
\text{(B)} & \text{NH}_2\text{--CH=C=CH}_2 & \quad\text{and}\quad \text{(C)} \quad \text{HC}{\equiv}\text{C--CH}_2\text{--NH}_2
\end{array}
$$

A canonical name could be constructed at this point, aligning the bond relations to give the smallest decimal number. (Atoms and their bond partners with lower numbers come first.) We can write

(A)   1 2 2 1 2 3 3 2 3 4 4 3
(B)   1 3 2 3 2 4 3 1 3 2 4 2
(C)   1 4 2 3 2 4 3 2 4 1 4 2

The canonical names already differ at the second position. For larger molecules, additional rules are included in practice (for example, the number of edges of a vertex in a hydrogen-pruned molecular graph, or the number of free electron pairs of an atom, or a hierarchy for certain atoms bonded to another atom of higher atomic number). Because of their completeness the published algorithms[27-29] have the important consequence of detecting constitutional symmetries of the molecular graph. The three hydrogens in a methyl group are always collected in an unpartitionable final set, meaning that their numbering scheme can be chosen arbitrarily. For 2-chloropropane, for example, the six methyl hydrogens and the two carbon atoms C_1 and C_3 form two final sets of constitutionally equivalent atoms that can be numbered freely. However, topological symmetries might not match symmetry groups of the 3-D molecular structures (or substructures, for local symmetries), and topologically equivalent atoms might not be such in a reactivity space or in a spectroscopic space.

4. Display of Two-Dimensional Molecular Models

Today the alphanumeric string input techniques are being replaced by the more user-friendly graphical input methods. These have the great advantage of being much faster, and they avoid coding errors due to the instantaneous vision of the generated structure. They allow a great deal of structural modification during an interactive input session. The chemist uses a light pen or a mouse to draw lines (representing bonds) on the screen of his graphics terminal. A number of click-on menus are used to define heteroatoms, double and triple bonds, formal charges, etc., thus turning the monochromatic molecular graph into a 2-D molecular model. Hand drawing normally is done rapidly and results in somewhat skewed structures. However, this has no effect on the internal representation, as no additional metric is included in the topological code of the molecule, still initially consisting of only a connectivity matrix.

Conversely, programs have been developed to plot 2-D molecular structures in a print-like quality starting from connectivity matrices. They represent an important communication channel to convey computer-generated chemical structures to the user's attention. They also can be used for redrawing handmade input structures to obtain customary, well-balanced structural images.

Attention must be focused on the following determinant points when designing a procedure for 2-D structural display:

1. Ring systems must be perceived and coordinates must be assigned that depict the structure in an easy-to-read way.
2. Bonds must be of reasonable size, avoiding excessive stretching when crowding occurs.
3. Coordinates of acyclic atoms should be assigned in a way that minimizes such crowding.
4. Complete structures should be oriented according to customary use; a steroid, for example, should be oriented with its A ring on the left and below the D ring of the tetracyclic system.
5. Similar structures should be plotted in a similar orientation.

This is a difficult task, and different solving approaches have been proposed. The system developed by the Chemical Abstracts Service[30] involves the use of a data base of ring templates. This library approach facilitates the uniform orientation of structures with common ring systems. Other approaches do not require substructural libraries. One general approach is designed to display 2-D structures for alphanumeric terminals and line printers. The procedure first constructs a 3-D model, rotates it in order to find the most planar view, and then projects it on two dimensions. However, the limited, hardware-dependent degrees of freedom in the choice of display coordinates make it less attractive than modern plot procedures designed for graphics terminals and plotters.[31]

A novel heuristic program to generate good display coordinates is based on the following six principal steps:[32]

1. Initial feature perception
2. Ring-system perception and assignment of relative coordinates to the ring atoms
3. Assignment of absolute plot coordinates to atoms
4. Coordinate refinement
5. Additional feature perception and structure manipulation
6. Plotting the final optimized structure

During the feature perception step, graph-invariant codes are generated for atoms, cycles, and ring systems. (Features independent of the connection table's sequence numbers are termed graph invariant.) These graph-invariant codes, which are a canonical numbering of the graph, are used (1) to minimize atom crowding and bond overlap when absolute coordinates are computed for all atoms, (2) to correctly orient ring systems, (3) to orient similar ring systems in a similar fashion, and (4) to obtain coordinates independent from sequence numbers.

Absolute coordinates are assigned to atoms breadth first by starting from an initial node. A procedure provides for exerting a repulsion between atoms. This artificial fleeing tendency is enhanced for cyclic atoms, atoms located near the center of the molecule (which show higher code indices than more distant ones), and atoms that are constitutionally crowded.

An energetic intervention of heuristic tools adapts the structure to its final plot version. Structural improvement is attained by rotating substituents, bending substituents, and slightly stretching acyclic bonds. Furthermore, a ring system is assigned a conventional orientation by rotating or flipping the whole system. Additional heuristic cosmetics refine the optical attractiveness of the pictures (e.g., the perceived aromatic nuclei are depicted with the familiar circle in the center of the polygon).

Heuristics do not guarantee completeness, and high-quality drawings cannot be produced for all possible structures. Experience, however, shows that most structures are displayed well and that intricate ring systems are represented in a way to raise an initial perplexity in the observer only in cases where any chosen perspective view would have no immediate interpretation. Figure 6 shows a number of paper plots of computer-generated 2-D molecular structures obtained with the described procedure.

Having discussed some of the principal topics of the coding and perception of topological molecular structures, their input, and their internal representation, we shall deal with what molecules truly are: three-dimensional entities.

C. GENERATION AND REPRESENTATION OF THREE-DIMENSIONAL MOLECULAR MODELS

The step from a 2-D to a 3-D molecular model in internal representation is short: a matrix $C(m,3)$ of the x,y,z coordinates of the m atoms in a molecule is generated and added to the topological information. The main question is how to obtain the Cartesian triples.

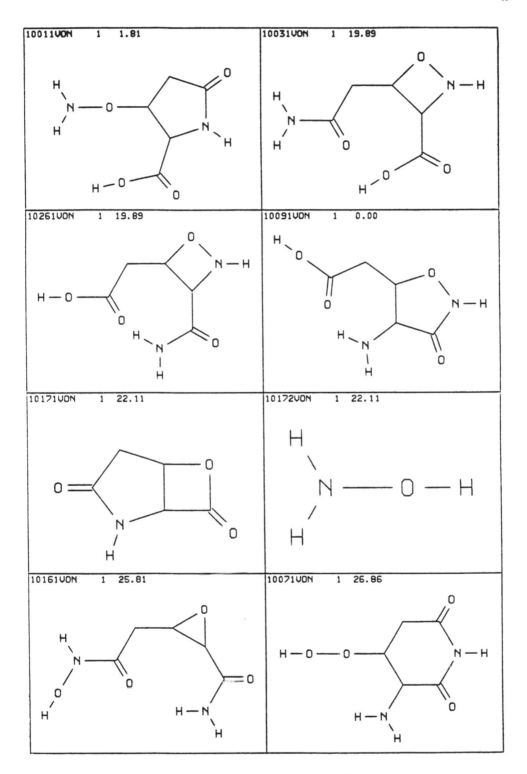

FIGURE 6. An example of computer-generated drawings of 2-D molecular structures.

1. Three-Dimensional Molecular Structures from Data Banks

The most direct way to obtain 3-D information about a known molecular structure is to access a crystallographic data bank, in which the X-ray crystal data of compounds are collected and from which they can be retrieved. Our knowledge about molecular geometry is mainly rooted in the results offered by X-ray crystallography. It is an accurate and increasingly fast analytical method which informs us about reliable interatomic distances, valence electron density, atomic thermal motion, and absolute configuration. Not only are small organic molecules the objects of diffractometry, but protein structures are also investigated, requiring an order of magnitude higher effort in the solution process.

The mightiest collection of structural and corollary data for organic, organometallic, and inorganic compounds is found in the files of the Cambridge Crystallographic Data Centre (Lensfield Road, Cambridge, CB2 1EW, U.K.). The files are available on magnetic tape and are divided (following a historical evolution) into three sections: the bibliographic section, the connectivity section, and the data section. For each retrieved structure one can interactively obtain

- The compound name
- Journal references and authors
- Stoichiometry data
- Information about physical form
- Cell geometry data
- Symmetry data
- Atomic coordinates

The crystal structures play one more important role when used to calibrate the functional parameters of the empirical (molecular mechanics) and semiempirical (MO methods) structure-generating methods. Precise vision of molecular geometries is a window to the study of structural changes arising from either (1) a chemical substitution at a specific position or (2) electronic inductive or mesomeric effects that can change internal angles or bond lengths. However, the atomic coordinates of a flexible molecule in a crystal represent atomic arrangements in the space under the influence of the crystal field. The adopted conformation is one of minimal strain energy in an environment of orderly arranged identical molecules. We must be aware that a molecule recalled from a crystal file, when visualized on a graphics screen, appears in this one particular conformational energy minimum. In other kinds of interactions (with solvent molecules, with a rigid receptor, or with reactants) the operating conformation might differ from the crystallographic one. Any further speculation can only follow the manipulation and physicochemical parameterization of the initial molecular model.

Data retrieval is good for current or retrospective information acquisition, but it is not always the method of choice for prospective modeling purposes, especially when novel families of structures for which crystal data are scarce or not yet available must be investigated. Due to their size (ca. 60 Mb for the Cambridge Files and 40 Mb for the Brookhaven National Laboratory Protein Data Bank [Upton, NY]), relevant mass storage capacity is required to install and maintain these files. It is not always possible to load a complete crystal file on the hard disks of small computers, leaving a telephone connection to public systems as the only (slow) way for an inexpensive data search.

2. Molecular Atomic Coordinates from Bond Parameters

A popular method of creating Cartesian 3-D coordinates for atoms in a molecule is their blockwise construction using internal parameters such as bond length and bond angle. This input style is still often used in quantum mechanical software and is available in many computer programs (e.g., ATCOOR[33]). The principle of calculation is simple and is based on trigonometric reasoning. It starts from an initial segment of three connected nonlinearly arranged atoms. The first atom is placed at the origin, the second (arbitrarily) on the positive y axis, and the third in the yz plane with a positive z coordinate.

The fixed coordinates of the first atom triple, A,B,C, with respect to the first atom A are given by the equations

$$x_A = y_A = z_A = x_B = z_B = x_C = 0 \tag{4a}$$

$$y_B = l_{AB}; \quad y_C = l_{AB} - l_{BC} \cos (ABC) \tag{4b}$$

$$z_C = l_{BC} \sin (ABC) \tag{4c}$$

The coordinates of a new atom D, attached to C, are calculated from its bond length (l_{CD}), its bond angle (BCD), and its dihedral angle $(ABCD)$ as follows:

$$x_D = -l_{CD} \sin (BCD) \sin (ABCD) \tag{5a}$$

$$y_D = l_{CD} \cos (BCD) \tag{5b}$$

$$z_D = l_{CD} \sin (ABC) \cos (ABCD) \tag{5c}$$

If more than four atoms are present, the procedure relocates atom B on the origin by a translocation (originally occupied by A), atoms C and D undergo a rotation similar to B and C in the first iteration, and the next atom, E, can be processed in analogy. Thus, the coordinates obtained by Equations 5a, 5b, and 5c can be understood as temporary coordinates for a general set of four atoms, O,P,Q,R. They are relative coordinates inside the frame set by the atom quadruple being processed, but must be transformed at the end into absolute coordinates in the primitive frame established by starting atom A. This can be achieved by adding two corrections to the temporary coordinates of a general atom R: a translational increment coming from the accumulated absolute coordinates x_O, y_O, and z_O of the actual atom O residing in the origin in a particular iteration, and a rotational term coming from the preorientation of atoms P and Q (exactly as B and C were preoriented at the beginning, reproducing Equations 4a, 4b, and 4c with only a change of subscripts A,B,C into O,P,Q).

The equations for the absolute coordinates of a general atom R belonging to quadruple O,P,Q,R, with O in the origin, P on the positive y axis, and Q in the yz plane, are given below. (For a complete derivation, see Reference 33; superscript T denotes temporary coordinates.)

$$x_R = x_O + {}^Tx_R[(\cos \alpha)(\cos \beta) + (\sin \alpha)(\sin \sigma)(\sin \beta)]$$
$$+ {}^Ty_R(\sin \alpha)(\cos \sigma) + {}^Tz_R[(\cos \alpha)(\sin \beta) - (\sin \alpha)(\sin \sigma)(\cos \beta)] \tag{6a}$$

$$y_R = y_O + {}^Tx_R[(\cos \alpha)(\sin \sigma)(\sin \beta) - (\sin \alpha)(\cos \beta)]$$
$$+ {}^Ty_R(\cos \alpha)(\cos \sigma) - {}^Tz_R[(\cos \alpha)(\sin \sigma)(\cos \beta) + (\sin \alpha)(\sin \beta)] \tag{6b}$$

$$z_R = z_O - {}^Tx_R(\cos \sigma)(\sin \beta) + {}^Ty_R(\sin \sigma) + {}^Tz_R(\cos \sigma)(\cos \beta) \tag{6d}$$

As described before, the molecule is rotated by α (around the z axis) and by σ (around the x axis) to bring P onto the positive y axis; then a rotation about the y axis by β brings atom Q into the yz plane with a positive z coordinate.

Again we have a situation with manual input of information provided by alphanumeric symbols. The method is revealed to be too clumsy for efficient and rapid modeling purposes. It assembles piecewise the smallest units contained in a molecule: the atoms. The connection rules are determined by simple geometric constraints: bond lengths and angles. A step toward an extension of this approach to a system that intelligently assembles building blocks larger than just atoms is intuitive.

3. Assembly of Structural Fragments

A useful way to construct 3-D molecular structures interactively via graphical input, realized in some molecular modeling systems, involves the use of the Cambridge Files. The files act as a library of known 3-D molecular structures. For example, the MMMS[34] and the CAMSEQ-II[35]/CHEMLAB* systems rely on this approach. Both allow the user to start with a structural fragment retrieved from the crystallographic data bank and to expand this substructure in the manner presented in the above section, i.e., from internal coordinates. Constructing polycyclic systems for which suitable fragments are not readily available in the library requires too much alphanumeric human intervention, however. Other systems are more generally adaptable to a larger variety of molecular structures and are completely independent from crystallographic structures. CHEMGRAF (The Lodge, Botley Works, Oxford OX20NN, U.K.), MMSX/SYBYL (Tripos Associates, 6548 Clayton Road, St. Louis, MO 63117), and SUPERNOVA-M (Tecnofarmaci SpA, Piazza Indipendenza, 00040 Pomezia [Rome], Italy), for example, all contain in their 3-D generating modules an extended set of structons, i.e., of elementary structural templates like simple atoms, functional groups (OH, SH, NH_2, CN, $COOH$, etc.), acyclic fragments, and ring subunits of various sizes. All of these structons can be joined together to form larger structons or complete molecules. They can be modified freely by adding or deleting bonds and atoms and by changing geometric parameters (e.g., bond lengths, dihedral angles). A ring-fusing option allows the user to fuse two rings along a bond.

The powerful MIMIC system (Molecular Graphics Laboratory for Organic Chemistry, University of Lund, Lund, Sweden) includes options for fusions along more than one bond — spirofusion and bridge building. The heart of the ring-generation module in MIMIC is RINGS.[36] It is a straight 3-D, visually controlled, interactive, and user-friendly menu-operated program. Ring systems are built by assembly of preconstructed ring templates of appropriate conformation (usually energetically preminimized). The stereochemistry of ring junctions is established by indicating the directions of the new bonds. The mechanism of joining structons to higher structures (cyclic or acyclic systems) is governed by preparing free valences, which represent the linking points of the structons. If, for example, one wants to generate toluene, the two predetermined structons benzene and methane will be recalled onto the screen. Then, to join CH_3- with C_6H_5- using the command JOIN, two hydrogen atoms (one of methane and one of benzene) are indicated by the operator. The corresponding C atoms are detected automatically by the program, which superimposes the marked bonds, deleting the H atoms and joining the two C atoms at a correct distance, thus yielding toluene. An identical procedure of connection holds for fusing two structons along one or more bond: the command FUSE expects two hydrogens per structon to be marked as future linkage positions. For example, two chair cyclohexane molecules give *cis*- and *trans*-decaline.

MIMIC, as well as the other systems mentioned previously, provides a useful interface for either (1) a subsequent molecular mechanics calculation aimed at establishing the lowest

---

* CHEMLAB, Molecular Design Ltd., 1122 B Street, Hayward, CA 94541.

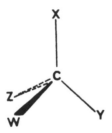

FIGURE 7.   A common wedged symbolic depiction of a chiral center.

energy conformation or (2) any other program designed for further manipulation of the generated molecular model.

4. Stereochemistry

The 3-D arrangement of atoms in a molecule exerts a dramatic influence on that molecule's behavior. We know that in natural products, as well as in synthetic biologically active compounds, stereochemical considerations play a dominant role. It is therefore mandatory that computer chemistry research also turns its attention to the automated perception and formal description of stereochemical relationships of atoms in a molecule. Modern computerized chemical documentation systems seem to follow this direction.

Stereochemical descriptors of organic molecules are of a binary nature, (R,S), (d,1), (erythro,threo), (cis,trans), and (E,Z) all corresponding to a binary (1,0) or parity ($+1, -1$) notation.

It would be useful if a computer would autodeductively attribute to each chiral center a dual description without human intervention, just by automated analysis of the 2-D symbolism of molecular structures. Most molecular graphics programs allow the user to define the stereochemistry around a certain atom by drawing bonds in wedged or hashed form (see Figure 7).

The determination of R or S configuration is sometimes difficult by visual inspection if the priority of the ligands is uncertain or complex to define. A simple case like the one in Figure 7 certainly does not need computer support, but for larger molecules containing many ambiguous chiral centers, human inference will become very slow and fallacious. A program has been developed that perceives the actual configuration of chiral centers and labels them with the traditional R/S symbolism starting from the user-input hashed/wedged molecular graph.[37] It should be noted that no coordinates are required for this task, as chirality (although dealing with the relative arrangement of atoms in space) is essentially nonmetric in the 3-D sense.

Establishing chiral descriptors requires the ranking of the ligands around the chiral center. In principle, any heuristic weighting method for ligands could attain a canonical numbering of ligands, but it seems useful to be consistent with the accepted descriptors R/S and E/Z in order to make global molecular names (stereochemical plus topological canonization), as introduced in chemical documentation systems, familiar to non-computer-oriented chemists, too. Therefore, in the following approach the ligand weighting and ranking procedures are based on the Cahn-Ingold-Prelog (CIP) rules.[50] The CIP rules originally were developed without regard to their algorithmic implementation, but after a reiterated revision a more rigorous formal frame could be found that allowed their computerization. Recently it has been proposed that hierarchical digraphs be spanned to label the priority of ligands.[51]

A hierarchical digraph of a stereogenous entity is a weighted, directed, polychromic, acyclic graph. It reflects the molecular connectivity, with the bonds (edges) protruding from a given initial node going in an outward direction to the ends of the generated branches.

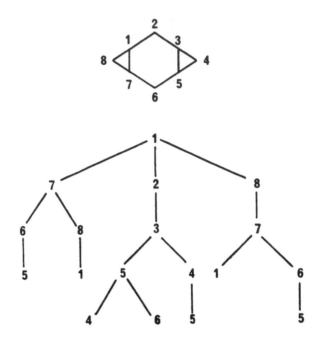

FIGURE 8. Example of a hierarchical digraph of tricy-clo(5.1.0.0$^{3.5}$)-octane spanned from atom 1.

The tree of Figure 8 is generated along shells. After each shell expansion a comparison of the weights of the included atoms is performed. The weighting follows the CIP rules, which attribute to each ligand node integer numbers obtained from the atomic core number. Other rule extensions (the inclusion of the atomic mass, for example) provide for increments or detractions in the ligand weight. From this canonization step four numbers, corresponding to the ligand weights, are obtained and ranked. Equal weights call for a further shell extension. The procedure can be summarized as follows: let **i** be a stereocenter marked by pseudostereo bonds or user-given stereo descriptors.

1. Atom **i** → root of the tree; **j** = 1

2. Generate tree shell **j**

3. Determine ligand weight W(k)

4. Compare ligand weights

5. Ligand weights equal?

6. Return vector of ranked ligands

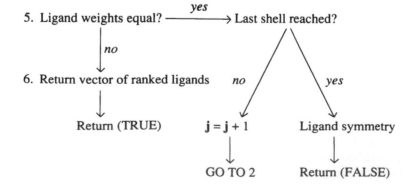

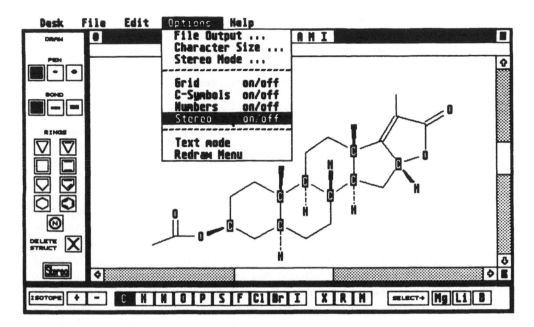

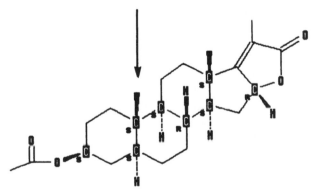

FIGURE 9. An off-screen plot of an automatic stereodescription for a complex structure. The stereocenters are labeled R and S.

An example of a computer-inferred determination of chirality descriptors is given in Figure 9. The menu-driven program first draws a 2-D model of the manually (by bond list or by mouse drawing) input structure on which the chiral centers are highlighted by the chemist with wedged bold and hashed bonds. Pressing the menu button **STEREO ON/OFF**, the analysis is performed and the molecule redrawn with the R/S labels properly placed at the chiral atoms.

Some considerations are necessary here: the automatic stereodescription as such is a step forward in the forging of global molecular canonical codes. The CIP rules, unfortunately, are still lacking the advantage of completeness, meaning that some structures with different ligands contain atoms which are *not* differentiated by the weighting rules accompanying hierarchical digraphs. Thus, we can regard the above program as a heuristic procedure which (due to the inherent CIP limitations) does not guarantee an unambiguous solution for all molecular structures.

5. Display Techniques of Three-Dimensional Molecular Models

Again we can say, "A picture tells you more than a thousand words (or numbers)." A touch of art certainly is included in the computer-generated graphical representation of 3-D molecular models. They range from the simple skeleton representation in black and white to more sophisticated color space-filling models of different styles. Some examples are shown in Figure 10 and in Plate 1.* A number of similar programs for representation of space-filling or CPK models are available (for example, see References 40 to 44).

6. Manipulation of Three-Dimensional Molecular Models

Interactive manipulation of 3-D models using computer graphics is a key technique used to look dynamically at a molecule and grasp its shape features. More than 99% of all chemistry students and researchers are continuously dealing with 2-D representations of molecules because that is how they find them printed in books.

The habit is to accept them as such, but the novelty of a fast and direct representation of more realistic 3-D models with a computer will increasingly kindle a different psychological perception of chemical structures, especially at an educational stage among students. Computer graphics tools for chemistry should be introduced more and more among faculties and should be freely accessible to chemistry students. The direct consequence of seeing things differently turns out to be a different way of thinking about them, and therefore different questions will surge in the researcher's mind, stimulating new answers. With the virtual, computer-generated structures displayed on a screen, the chemist can perform operations which are not feasible with material models. Besides the trivial operation of rotating the whole structure or selected substructures around a given bond, one major application is the superposition of rigid structures to inspect shape differences. Comparing structures can, for example, unveil the distortions introduced in a parent molecular frame by chemical substitution or quantify the differences in shape among various compounds belonging to a single class (or different classes) of chemical or biochemical behavior; the shape differences (together with other parameters, of course) can then be used to rationalize the relationship between chemical activity and molecular structure in a specific environment. The superposition of molecules is a general tool available today in all major molecular modeling systems. It is an interactive approach in which the user selects a set of n atoms (a substructure if they are connected) of a molecule M to be superposed and then relates them to n fixed atoms of a second set belonging to the reference structure R. The matching of the two selected sets of atoms requires the sum of the squares of the euclidean distances between related atoms to be minimized. Written out formally,

$$\sum_i (\mathbf{r}_{Mi} - \mathbf{r}_{Ri})^2 = \text{minimum} \qquad (i = 1,n)$$

with $\mathbf{r}$ representing the position vector of an atom i in 3-D euclidean space. The minimization is done either using an algebraic solution method or by iterative numerical search for the best least-squares fit. If substructures with different numbers of atoms have to be matched, an atom may be related to a so-called superatom, which is the center of coordinates of a chosen set of atoms. (For example, pyridine and pyrrole can be represented by two superatoms.) Structure comparisons by matching are very popular in drug design research, especially in the search for pharmacophoric substructures and in receptor modeling.[47-49]

We shall illustrate the utility of computer manipulation of molecular models with two examples. The first example shows the difference in shape between two steroids, testosterone and estradiol, which carry opposite chemical messages in their biological activity: the first accounts for secondary male characteristics and the latter for secondary female characteristics. Inspection of the separate structures (Figure 11) shows that the D rings are similarly shaped

* Plate 1 follows page 62.

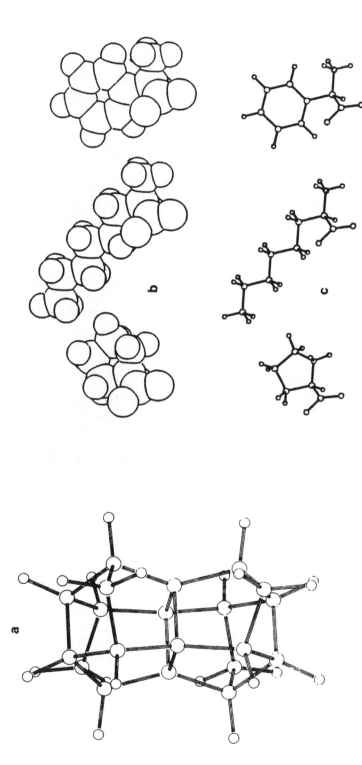

FIGURE 10. Different styles used to plot 3-D molecular models. (A) Ball and stick representation of pagodane (obtained with the program Marilyn).[45] (B) Van der Waals space-filling models and (C) ball and stick models of *R*-proline, *R*-phenylglycine and 2-*R*-aminooctanoic acid.

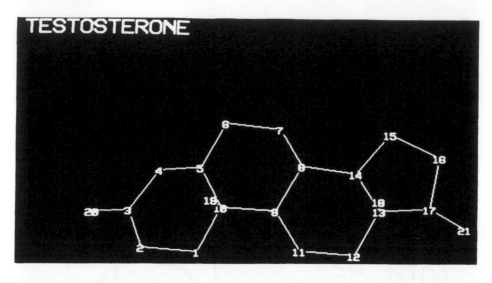

A

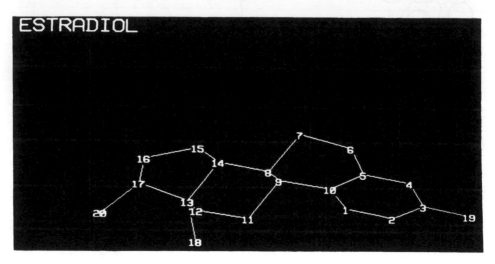

B

FIGURE 11.   The steroids testosterone (A) and estradiol (B) are shown on a graphic screen with numbered atoms. Atoms belonging to the D rings are selected for a subsequent superposition.

in both hormones. These common substructures are matched and the resulting ensemble of superposed structures visualized. The "side" view (Figure 12) very clearly conveys information about the spatial differences. The distance between the phenolic oxygen (estradiol) and the carbonylic oxygen at C-3 (testosterone) is 3.27 Å, and it is evident that the specific biochemical features inherent in the two steroids must be "coded" in their A and B regions.

The next example deals with the comparison of acetyl-β-methylcholine and natural muscarine. It is known that muscarine, a poison to man, blocks acetylcholinesterase, which is the enzyme responsible for metabolizing acetylcholine. Acetylcholine acts as a neurotransmitter in the body and is liberated at the synapses of nerve endings, which are physically separated from the dendrites of neighboring nerve cells. The neurotransmitter carries the impulse across this anatomical gap to the next nerve segment. For an orderly flow of electrical impulses to occur from autonomic nerve to nerve or from nerve to muscle, the chemical

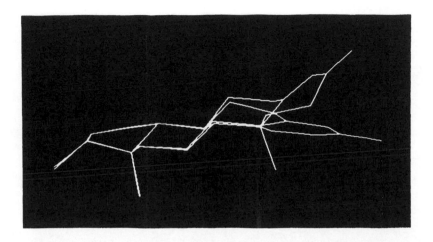

FIGURE 12. A side view of the two steroids with superimposed common substructures (the respective D rings).

mediator must be destroyed immediately after reaction with its receptor, causing a subsequent dissociation of the enzyme-transmitter complex. Thus, the receptor site is cleared for a new arriving neurotransmitter molecule. The question that interests us is the following: how can molecular modeling techniques provide evidence that muscarine shows a much higher affinity for the enzyme receptor, in an almost irreversible way? To show affinity means to be able to reproduce the geometrical and chemical features of the reference structure to a large extent. Figures 13A and B show the structures of muscarine and methylcholine as they were created by a 3-D structure generator. The common substructure in both molecules is the sequence N(+)–C–C–O–C–C. In muscarine part of this substructure is flexible, while in methyl acetylcholine the complete substructure is flexible (degrees of freedom in the dihedral angles).

To calculate the best match between a flexible molecule and a rigid reference structure, a program called PYTHON[50] has been generated as a natural extension of the DRACO system,[51,52] which was originally capable of superposition of rigid substructures only. It works on principles of molecular dynamics (see later sections) and acts on the dihedral angles of the flexible substructure to minimize the squares of the distances between matching atoms. The result of the fitting process is illustrated in Plate 2.* We can see that the ring oxygen of muscarine completely overlaps with the carboxylic oxygen of the acetylcholine derivative, and the carbonylic oxygen in this particular best-fit conformation comes very close to the hydroxyl oxygen of muscarine (ca. 1.5 Å). An interesting visual technique to qualitatively evaluate the degree of similarity and of interpenetration of two van der Waals models is offered by molecular tomography.[51,52] If van der Waals spheres are spanned around each atom of the superposed structures, a cut somewhere through the global molecular ensemble can be imagined. When slicing an orange, the cut gives insight into the internal structure of the fruit at different distances from its center. Cutting far from the core of the orange will generate slices with small radii, while cuts at or around the center will give slices with large radii. A tomographical view of superposed structures can therefore allow the visual perception of common regions in space, i.e., regions occupied by both molecules at the same time. In addition, the space regions occupied by only one species, the residual regions, are also visible. Color graphics make the visual perception very straightforward and user friendly. Red encodes the common regions, while blue and yellow reproduce each

*   Plate 2 follows page 62.

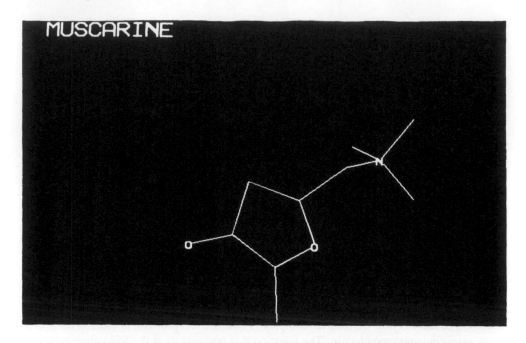

A

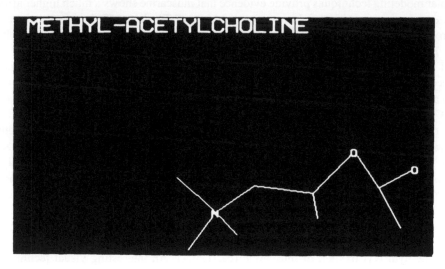

B

FIGURE 13.   (A) The structure of natural muscarine. (B) The structure of β-methyl-acetylcholine.

separate van der Waals molecular model. The red lens-shaped area between the two mentioned oxygens shows that there is a high degree of overlap between the two atoms: it seems
that the muscarine hydroxylic oxygen can take over the role of the strongly overlapping
acetylcholine oxygen in this particular conformation. An interpretation of the result given
by a combination of computerized molecular structure manipulation and computer graphics
helps us in rationalizing the probable reasons why muscarine has such a high affinity for
the acetylcholinesterase receptor: it mimics to a large extent the molecular shape and the
charge distribution of the natural neurotransmitter.

Going back to the superposed steroids, in analogy we can perform a tomographical cut to obtain a side view of the common and residual regions, which give an immediate perception of the shape similarities and dissimilarities of the two hormones (Plate 3).*

The concepts of shape similarity and of common and residual regions can be treated formally and quantitatively by Boolean mathematics and will be introduced in a later section.

II. GENERATION OF PHYSICOCHEMICAL PARAMETERS BY MOLECULAR MODELING TECHNIQUES

A. MOLECULAR VOLUMES, MOLECULAR SURFACE AREAS, AND SHAPE SIMILARITY

1. Boolean Encoding of Three-Dimensional Space-Filling Molecular Models

The quantitative evaluation of geometric parameters of molecules is unavoidable in reliable drug design studies. The molecular shape is one of the physical molecular aspects correlated with biological activity.[53-60] In chromatography[61] and solvation,[62] molecular volumes and molecular surface areas are important quantities included in the rational interpretation of these measurable molecular physicochemical phenomena.

So far we have dealt with molecular models represented by points (the atoms) in 3-D euclidean space. All of these points lacked extension. One should always be aware of the trivial though sometimes forgotten fact that any atomic x,y,z coordinate triple describes the mean location of the atomic core, but that chemistry involves the electrons, which belong to portions of space quite a distance away from the nuclei. A molecular skeleton therefore cannot properly represent the shape of a molecule and its space-filling properties.

Molecules, due to their electron density distribution, are better described as entities with a finite extension in physical space, i.e., as solid bodies. Recently a method was presented for the superposition of pseudoelectron density maps of molecules together with a definition of "excluded" and "essential" volume density maps.[63] Partial spatial aspects also have been described quantitatively by various authors, with the introduction of steric parameters[64] and of a molecular shape analysis.[65]

A computerized binary representation of a solid molecular body was developed[51,52,66] to allow both the computation of molecular volumes and surface areas and the calculation of an empirical degree of molecular shape similarity. We want to discuss this particular approach because it further demonstrates the applicability and the relevance of Boolean operators in chemical problem solving.

A portion of physical space is described by a tensor $\Gamma(m,m,m)$ of rank 3 having m components per axis. The space included by Γ is imagined as a virtual cube subdivided into m^3 subspaces, each of them characterized by an index triple (i,k,l). A properly scaled van der Waals molecular model can be embedded inside the cube given by Γ. All of the subspaces located inside the van der Waals model are labeled "1", while those outside are labeled "0" (see Figure 14).

This is a purely logical Boolean description of the pattern of distributed solid matter in 3-D space. In a computer representation, the 1s and the 0s are equivalent to TRUE and FALSE, which are logical variables. Note that this method does not merely encode the outer shell (the peel) of the molecular model, but also its interior as a matter-filled solid body. To provide for good resolution of the molecular shape, a minimum of about 1 million subspaces in the virtual cube seems necessary.

In real programming work the virtual cube must be represented by the tensor $\Gamma\,(m,m,m)$, in which the m components per axis can conveniently be chosen coherent with the CPU architecture. For example, to save core memory, a compact bit-to-bit encoding may be preferred to the use of simple logical variables or even integer numbers. If one works on a

* Plate 3 follows page 62.

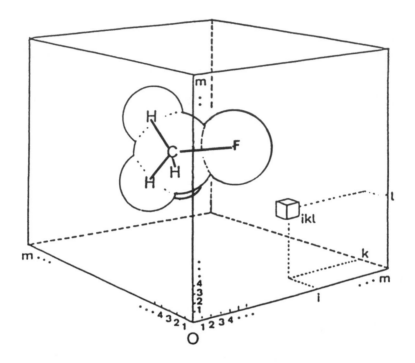

FIGURE 14.   A schematic representation of the virtual cube (a Boolean tensor) in which a scaled van der Waals molecular model is imbedded. The cube is subdivided into elementary subspaces (i,k,l) that can have the values 1 or 0, depending on whether they reside inside or outside the spatial range of the molecular model, respectively.

32-bit machine, a straightforward way to obtain about 1 million subspaces is to generate them from an equivalent number of bits in a program-internal tensor $\Gamma_P = \Gamma_P(96,96,3)$, in which we have for the tensor components 96 words each for the x and the y axis and three words (32 bits each) spanning the 96 components along the z axis. Compared to an integer number codification, this technique saves core memory by a factor of 32. Special bit-addressing routines are then responsible for switching on and off a specific bit inside a certain word in Γ_P.

2. Boolean Tensor Operations

For each molecular model a specific bit tensor can be generated. These tensors can be handled easily by Boolean operators and deliver some quantities useful in molecular modeling. The molecular volume V is obtained by summing all TRUE bits $(ikl)_\tau$ inside the virtual cube:

$$V = \sum_\tau (ikl)_\tau \tag{7}$$

The molecular area is calculated by the sum of all TRUE bits that are placed next to any FALSE bit, the latter evidently defining the outer bit shell at the van der Waals border of the molecular model.

The Boolean combination of two (or more) bit tensors is also a bit tensor. The resulting bit patterns quantitatively describe special space regions related to the common and residual regions introduced qualitatively with the molecular tomography above. Resuming the action of molecular superposition, we can define the following molecular body regions derived

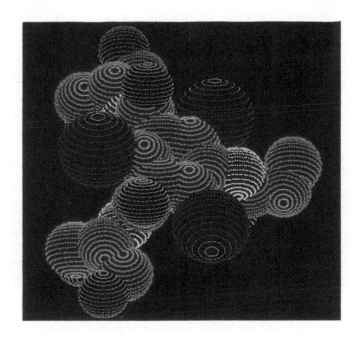

A

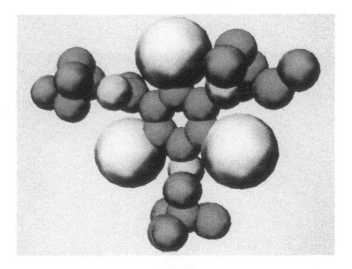

B

PLATE 1. Colors bring more visual information. (A) A three-dimensional model of iopamidol with dotted contour lines of the van der Waals spheres for the atoms iodine (violet), oxygen (red), nitrogen (green), and carbon (blue). (B) The same molecule is shown in a different color map as a shaded solid three-dimensional van der Waals model (iodine in yellow).[34]

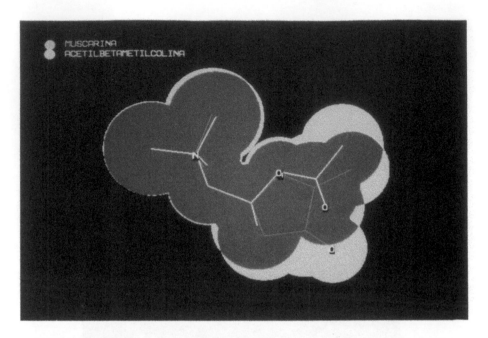

PLATE 2.   Superposition of the common substructures N(+)–C–C–O–C–C and automatic fitting of
the dihedral angles of the flexible acetylcholine derivative to the geometry given by the rigid muscarine
structure leads to a maximum structural overlap. A tomographical view visualizes common regions in
space (red area) and the residual regions occupied only by muscarine (yellow area) and only by β-
methyl acetylcholine (blue area). The carbonyl oxygen of the latter species invades to a large extent the
space regions of the fixed hydroxylic oxygen of muscarine.

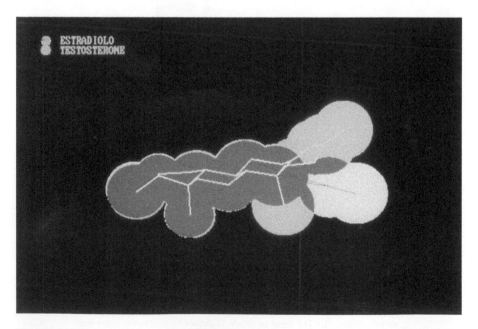

PLATE 3.   A tomographical view of the superposed hormones testosterone and estradiol. Common
and residual regions are clearly visible through the color mapping. The largest shape difference between
testosterone and estradiol is in the region of their respective A and B rings.

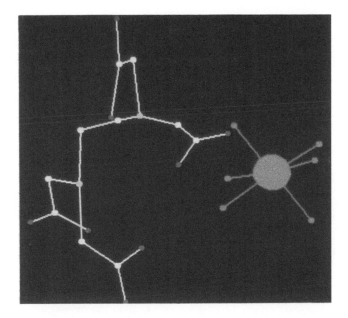

A

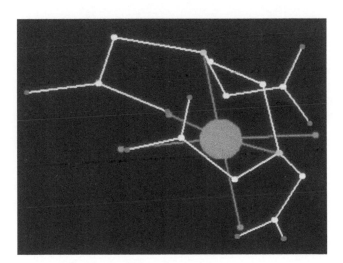

B

PLATE 4. (A) The Ni(II) virtual coordination octaheder and the EDTA molecule shown separated in three-dimensional space (red = oxygen, green = nitrogen). (B) Matching of the six donor atoms of the flexible ligand to the fixed coordination points with molecular dynamics techniques leads to the simulated formation of an Ni(II)-EDTA complex.

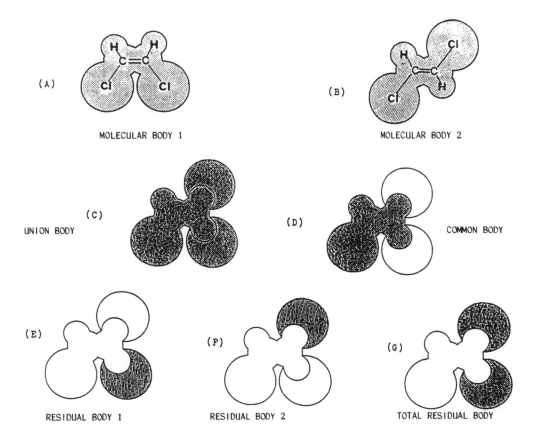

FIGURE 15. Two molecular bodies are superposed according to their common substructure. The five secondary bodies (C, D, E, F, and G) are the result of logical operations performed on the two Boolean tensors encoding the two space-filling solid molecular models (A and B).

from the spatial overlap of two virtual van der Waals molecular models, as illustrated with two generic molecular models in Figure 15.

The union of a first bit tensor (Γ_1) encoding molecule A with a second tensor (Γ_2) encoding molecule B results in a new tensor (Γ_U) encoding the union body, or superbody, of the two molecules in the chosen specific superposition:

$$\Gamma_U = \Gamma_1 \text{ OR } \Gamma_2 \tag{8}$$

The space region belonging to both superposed molecules is the common body, given by Γ_C:

$$\Gamma_C = \Gamma_1 \text{ AND } \Gamma_2 \tag{9}$$

The residual body, expressed by Γ_R, is obtained by subtracting the common body from the union body:

$$\Gamma_R = \Gamma_U \text{ AND NOT } \Gamma_C \tag{10}$$

This equation can be transformed into Equation 11 using Γ_1 and Γ_2:

$$\Gamma_R = [\Gamma_1 \text{ AND NOT } (\Gamma_1 \text{ AND } \Gamma_2)] \text{ OR } [\Gamma_2 \text{ AND NOT } (\Gamma_1 \text{ AND } \Gamma_2)] \tag{11}$$

The first term in Equation 11 encodes the residual body of the first molecule, the second term the residual body of the second molecule. Using the above bit-counting method the volumes V_U, V_C, and V_R of these special bodies are evaluated quantitatively. The extreme case of an isomorphic superposition of two identical molecules (such that V_R is zero) gives

$$V_U = V_C$$

and therefore

$$V_C / V_U = 1$$

For every other case

$$V_C / V_U < 1$$

A shape similarity index S can now be defined as

$$S = V_C / V_U \tag{12}$$

In the superposition examples of the previous section, the computation gives $S = 0.61$ for the steroids and $S = 0.75$ for the acetylcholine-muscarine study.

The reiterated superposition of molecular models by matching, for example, the largest common substructure is a method used in drug design research when trying to identify the shape requirements necessary for a postulated drug to be active in a specific biological environment. The generation of a superbody containing the shapes of several comparably active drugs can (1) give hints on the complementary shape of the receptor and (2) allow preclassification of new, postulated structures as either "probably active" or "probably inactive" according to their similarity to the superbody of the active species. This reasoning does not necessarily have to include the entire molecular structure, but can be limited to strategic parts of it.

The sequence of actions described for shape comparison can be applied only to certain portions of the molecular structure(s) (e.g., the suspected pharmacophoric regions) in order to determine the essential morphology of a specific pharmacophoric pattern. The main advantage of Boolean encoding is its simplicity of generation and manipulation. Other methods based on difficult dissections and geometric parameterizations of the molecular model have been developed to compute surface areas and molecular volumes,[66,67] but they do not provide a code for the molecular body. If required, the bit code can be canonized by aligning the molecule with its three moments of inertia along the tensor axes, placing the molecular center of gravity in the center of the cube.

The computed volumes have been successfully correlated with volume-dependent physicochemical parameters like the vaporization enthalpy[51,52] and the boiling point[68] of homologous organic compounds.

B. MOLECULAR ENERGETICS
1. Introduction
Molecular modeling techniques must provide both visual and numerical information, which (if properly combined) contribute to enhance the knowledge of the studied system. Speaking of "numerical" information activates the second mode of application of computers in chemistry, the numerical mode. It adds hard numbers to the more semantic symbol-interpreting and symbol-manipulating approach. One of these numbers is the molecular energy.

The difference in the role and quality of results of numerical programs today from results obtainable in past decades is found precisely in two aspects of changing chemical science: (1) the availability of much faster and less costly computers and (2) the combination of these purely numerical results with results coming from a semantic solution to chemical problems. Any better experimental strategy and deeper knowledge of the investigated system, any refinement in interpretation can only derive from a combined inference of both approaches. A single conformational energy value taken detached from all other possible conformation minima (i.e., a single problem state disconnected from the others surrounding it) has little practical meaning; in the same way, the generation of all possible isomers of a certain compound has no practical meaning if it is not supported by ring-strain energy and bond energy evaluations, which are necessary to eliminate "impossible" structures. Better computers and the parallel progress in computer chemistry science today allow, for example, the calculation of the entire conformational energy hypersurface of an organic molecule, even generating a good approximation of the Boltzmann distribution function with its corresponding populations. Furthermore, through the availability of the energy hypersurface we acquire information about the global behavior of the molecule. The reader will have noticed immediately that this is equivalent to generating all problem states around a certain initial node and linking them mutually by transition paths, the edges! Thus, information about the system as a whole, and not just about one single state, is gained: indeed a strategic achievement.

We see clearly that the two historically divergent computer-assisted chemical philosophies, the numerical and the symbolic/semantic ones, are converging here to become allies for a better overall research optimization.

In most cases interpretation of chemical phenomena, especially with organic compounds, requires the availability of some information on energy. One type is certainly the internal energy of a molecule. The internal energy can be split into two parts, one accounting for the formation enthalpy of an ideal molecule, the other for its steric energy. Each of these quantities is frequently put into a relationship with molecular geometry, reactivity, and pharmacological behavior. To obtain such an important characterization of a molecular structure, molecular mechanics (MM) calculations were originally introduced as computationally fast procedures for quantitative conformational analysis. (For a review of molecular mechanics, see References 69 to 74.) They are used today for the generation and refinement of molecular geometries (model builders, in the slang of computer chemists), for the rationalization of molecular vibrational properties, and for the computation of relative heats of formation and relative strain energies.

They are, however, of no utility when electronic effects are the primary conformation determinants; in such cases, quantum-mechanical treatments via molecular orbital calculations (MO) are far better for the prediction of molecular properties, at the cost of extremely long computation times. Thus, MM and MO methods should be seen as complementary rather than competitive techniques.

2. Molecular Mechanics: Empirical Force-Field Calculations

From the point of view of molecular mechanics, a molecule is considered to be an ensemble of atoms held together by classical forces, the force field, like spheres connected by springs. Energy differences between molecular species are estimated by classical mechanics, avoiding all complications arising from quantum-mechanical treatments. Being to a high degree an empirical approach it needs an extended parameterization, which restricts its use in the evaluation of relative quantities like the difference in strain energy between two conformers. Determination of absolute quantities like total molecular energy can be achieved only through an empirical adjustment of parameters to fit experimental values.

The current use of force-field calculations is showing a major shift toward computation of conformational strain, a fact greatly influenced by the present trend, which is oriented strongly toward structure-activity research in drug design. Generally, we can say that programs relating molecular geometries to strain energies are *de facto* minimization programs used to find the "most stable conformation". Many users of commercially available force-field programs tend to express a strong belief that the most stable conformation should be used in rationalizing pharmacological activity in drug design. However, most programs do not guarantee that the detected minimum *is* the absolute minimum, as any localized minimum depends on the starting situation of the internal molecular coordinates. Also, if the absolute minimum is found, no evidence so far can prove that it is really responsible for molecular reactivity. We think, on the contrary, that because of the interaction with a certain reactant (be it a biocatalyst or some small organic molecule) in the first stages of a chemical transformation, the molecule in question will not be in its absolute conformational minimum (with the obvious exception of heavily rotationally hindered structures). The reactivity depends not only on the energetic difference between educt and product, but also on its kinetic possibilities, which are related to the height of the activation barrier. The higher the educt is in energy (that is, the farther away from the absolute conformational minimum), the shorter the climb over the barrier and, therefore, the faster the reaction. Again, this reasoning holds for situations in which electronic effects do not play the dominant role, which can in turn be so energy demanding or orientation controlling that the conformational terms, normally smaller in magnitude, become negligible.

Molecular mechanics is also a valuable tool for inspection of molecular properties, and it has given rise (as discussed below) to molecular dynamics methodologies which seem to obviate some of the shortcomings of static molecular mechanics.

The central concept in MM programs is strain. It is an inexact but accepted intuitive feature among chemists. Strain-free molecules are regarded as ideal, whereas strained conformations, if somehow allowed by molecular flexibility, are bound to higher energies. The energy expression is dependent on geometric parameters like bond angles (Θ), bond distances (r), torsional eclipsing angles (Φ), and nonbonded distances (d).

The total strain energy (E_T) can be formulated empirically as

$$E_T = E(r) + E(\Theta) + E(d) + E(\Phi) \tag{13}$$

The single contributions are as follows:

$$E(r) = \Sigma\, k_r(r - r_0)^2 \tag{14}$$

$$E(\Theta) = \Sigma\, k_\Theta(\Theta - \Theta_0)^2 \tag{15}$$

the summation going over all bonds and bond angles, respectively.

Hook's law is only a good approximation for small displacements from the undeformed ideal situation at r_0 and Θ_0. For larger deformations higher power terms are added, and Morse potentials are substituted for stretching. One must keep in mind that increasing the number of functional terms increases the quality of the reproduced properties, but requires the inclusion of additional adjustable parameters (the force constants), a mechanism that obscures the intuitive interpretability offered by Equation 13.

In addition to stretching and bending terms, a contribution coming from nonbonded van der Waals interactions between all pairs of atoms neither bonded to one another nor to a common atom is given by the expressions

$$E(d) = \Sigma\, a_p d^{-12} - \Sigma\, b_p d^{-6} \qquad \text{(Lennard-Jones potential)} \tag{16}$$

$$E(d) = \Sigma\, a_p \exp(-b_p d) - \Sigma\, c_p d^{-6} \qquad \text{(Buckingham potential)} \qquad (17)$$

where subscript p indicates the pth kind of atom pair. The adjustable parameters a_p, b_p, and c_p may be evaluated from experimentally determined physical properties and are constants specific for each pair (p) of atom types (carbon-carbon, carbon-hydrogen, etc.). Both potentials assume a d^{-6} attractive component, which is the induced dipole/induced dipole term in the multipole expansion of the dispersion energy between two polarizable systems. The repulsive contribution in the nonbonded Lennard-Jones potential appears as d^{-12}, which is taken mainly for mathematical convenience, although other powers of d seem more realistic. The Buckingham repulsive term is exponential and provides for stronger repulsion.

If the calculation of the conformational energy only involved the three contributions of stretching, bending, and van der Waals forces, one would find the eclipsed form in ethane to be insufficiently less stable (i.e., higher in energy) than the staggered conformation. To compute a more realistic value for eclipsed arrangements of atoms, special torsional terms are included in the overall energetic function, justifying the fourth potential in Equation 13. The expression for $E(\Phi)$ is given by the equation

$$E(\Phi) = \Sigma\, V_\phi\,[1 + s(\cos n\Phi)] \qquad (18)$$

where V_ϕ is the barrier of free rotation with periodicity n and s is a parity counter for a staggered minimum ($s = 1$) or an eclipsed minimum ($s = -1$)

The selection of modified potentials introduced, for example, by cross-terms reproducing bend-torsion-bend displacements, by special equations modeling the energy of hydrogen bonding, by Coulombic potentials, or by the power expansion of stretching and bending potentials has been proposed as well. The question of whether to use Cartesian or internal coordinates has been the object of much study in the past. These particular aspects have been discussed in review articles[70] and are not treated further here. Whatever degree of sophistication is chosen, the basic conceptual role of MM does not change within a general view of computer chemistry as understood here.

All force-field programs have one thing in common: they consist of multivariate functions that must be minimized. These functions in turn include a large number of parameters that must be adjusted to reproduce experimental evidence. The latter problem is solved by either (1) trial and error adjustment of force constants and reference geometries in order to obtain the best possible fit between calculated and observed properties or (2) least-squares fitting of the adjustable parameters to experimental findings. The former problem involves the choice of some minimization routine, always being aware that an unlucky initial arrangement of the atomic coordinates may lead to a side minimum which can be distant from the absolute lowest energy extremum. A short overview of some common minimization algorithms will be presented here for the sake of completeness, keeping in mind that these methods are not "better" or "worse", but simply different mathematical techniques that can prevail depending on the degree of sophistication of the program, the computer speed available, and the kind of molecules processed (the number of atoms, the intrinsic shape of the conformational hypersurface).[75]

The first general scheme for finding an energy minimum was the steepest descent method.[75] The energy of a molecule is calculated with the coordinates corresponding to an initial trial geometry. A given atom i is then displaced by some amount Δx along the x axis and the energy recalculated. The same atom i is then shifted by Δy along the y axis and by Δz along the z axis, each time the energy being calculated anew. The procedure reiterates this mechanism for all other atoms j, k, l, . . . , and z in the molecule. The atoms are then moved simultaneously in directions that cause the highest reduction in energy. After one global iteration, which corresponds to positioning all atoms on new coordinates originating from the corrections Δx, Δy, and Δz, the energy is calculated again. If it is equal to the result of the previous step a minimum is assumed, but if it decreases other small displacements are introduced and the whole process repeated. An advantage of this method is that it appears to be free from stalling, never leaving a molecular geometry on a saddle point. Minimization proceeds rapidly in regions far from the minimum (steep regions of the hypersurface) and slows down in its proximity (the hypersurface becomes flat, the displacements smaller). A more sophisticated family of minimization methods consists of modifications of the general Newton iteration.[75,76]

The conditions for an extremal point are given by the equation

$$F'(\mathbf{x}^m) = 0 \tag{19}$$

where $F(\mathbf{x}) = F(x_1, x_2, x_3, \ldots, x_n)$ and $F'(\mathbf{x})$ is the vector of the first derivatives $\partial F / \partial x_i$. If $\mathbf{x}$ is a vector near a true minimum $\mathbf{x}^m$, let $\mathbf{x} + \delta\mathbf{x}$ be a better approximation than $\mathbf{x}$ alone. We can write

$$F'(\mathbf{x} + \delta\mathbf{x}) = 0 \tag{20}$$

and by the Taylor expansion

$$F'(\mathbf{x} + \delta\mathbf{x}) = F'(\mathbf{x}) + F''(\mathbf{x})\delta\mathbf{x} \tag{21}$$

Truncating the series after the linear term and insertion of Equation 20 into Equation 21 gives

$$F'(\mathbf{x}) + F''(\mathbf{x})\delta\mathbf{x} = 0 \tag{22}$$

and, finally,

$$\delta\mathbf{x} = -F''(\mathbf{x})^{-1}F'(\mathbf{x}) \tag{23}$$

Different implementations of Equation 23 have been developed, each with specific advantages and limitations. Also, the question of whether to use numerical or analytical derivatives has been treated in detail.[70] The general Newton iteration has found widespread application in the Newton-Raphson procedure and in many other related quasi-Newton modifications. The Newton-Raphson iteration can be formulated as follows: being that

$$\delta\mathbf{x} = \mathbf{x}_{k+1} - \mathbf{x}_k$$

for iteration $k \rightarrow k + 1$ it follows that

$$\mathbf{x}_{k+1} = \mathbf{x}_k - \alpha F''(\mathbf{x})^{-1}F'(\mathbf{x}) \tag{24}$$

The step length is $0 \geqslant \alpha \geqslant 1$.

Several variations of the basic iteration are known:

1. Steepest descent method

$$F''(\mathbf{x}) = \beta_i, \quad i = 1, 3n$$

2. Pure diagonal method

$$F''(\mathbf{x}) = (\partial^2/\partial x_i^2), \quad i = 1,3n$$

3. Block diagonal method

$$F''(\mathbf{x}) = (\partial^2/\partial x_i \partial x_j), \quad i = 3m + 1, \quad 3m + 3$$
$$j = 3m + 1, \quad 3m + 3$$
$$m = 0, \quad n - 1$$

4. Full matrix method

$$F''(\mathbf{x}) = (\partial^2/\partial x_i \partial x_j), \quad i,j = 1,3n$$

The better F'' approximates the full matrix Newton-Raphson method, the faster the iteration converges, but the more computational steps are required. MM practitioners usually find a compromise between CPU time requirements and minimization of speed. Experience shows that starting methods should also be able to deal with poor initial trial geometries, a situation where full matrix methods tend to fail in convergence. Therefore, the use of a preoptimization procedure that can transform, for example, a 2-D molecular structure depicted on a graphics screen into a chemically more reasonable 3-D geometry is an effective approach (e.g., steepest descent, simplex search).

Having disposed of the necessary mathematical introduction to MM, we may return to the essence of our general *leitmotiv*. In order to evaluate which aspects of MM can be relevant in computer chemistry, we must first understand the very meaning of an MM calculation, which is useful only if it can predict possible conformational behaviors of a molecule. Every single piece of information obtained by data processing with computer chemistry tools (we shall see in later chapters how such programs tend to produce information rather than data) serves to increase our chemical knowledge and, consequently, to allow better predictions for future experiments (i.e., to optimize experimental strategies). Now, programs that yield a single number seldom possess direct strategic capabilities, as the calculated numbers are frequently isolated in the space of all other possibilities conceivable in principle, but not yet generated for whatever reasons. These numbers, then, are just data in the mind of the chemist who has to attempt a hazardous extrapolation from data to knowledge. The point is that data are often mistaken for information. Information, in our view, is a rational structuring of data by which they become described through either some mathematical/logical function or some collective symbolic/semantic expression. The function is one possible way of expressing information, as it permits predictions of untested situations.

Coming back to the conformational aspect, energy data for different conformations (if modeled by some function) generate information, possibly the complete information about the absolute minimum and all the relative energy minima of a flexible molecule together with the energies of the transition barriers. The conformational energy hypersurface is the information, which can be used to gain knowledge about the chemical and physical behavior of the structure under changing conditions.

A meritorious effort was recently made in this direction to solve a fundamental problem of conformational analysis,[77-79] it deserves to be mentioned here.

Let a conformational state S_i be described by a set of n dihedral angles $\{\Phi_1, \Phi_2, \Phi_3, \ldots, \Phi_n\}$. For the conformational energy we can write

$$E_S = F(\Phi_1, \Phi_2, \Phi_3, \ldots, \Phi_n) \tag{25}$$

where bond angles and bond distances are kept constant in a preminimized optimized geometry.

The conformational space has an infinite density of problem states, as the energy hypersurface is a continuum. Therefore, a discretization of the space is accomplished by allowing the dihedral angles to vary after some given increment $\Delta\Phi$, which is chosen small enough so as to reliably model the shape of the multidimensional E_s function.

The existence probability $p(S_i)$ of S_i depends on E_s through the Boltzmann statistics

$$p(S_i) = \exp[-E_s/kT]/Q \tag{26}$$

where k is the Boltzmann constant, T the absolute temperature, and Q the partition function over all states;

$$Q = \int \exp[-E_s/kT]dE \tag{27}$$

The integral is extended over all possible conformational states, but calculations show that only a fraction of them have nonnegligible probabilities of existence. After choosing an initial conformation, the dihedral angles are modified stepwise by a given increment. Systematic checking of each incremented conformation to individuate unacceptable contacts between atoms reduces the total number of minimizations to be performed,[79] and for each new conformational state the energy is calculated in the usual way if one of the following conditions holds:

1.   The difference between E_s of the current $S(k)$ and the lowest minimum found so far is not smaller than some fixed threshold.
2.   The distance of the current point from an already established minimum is not larger than the average value of distances between that minimum and the nearest saddle points along the n different Φ axes.

The generation of conformational states ends when all combinations of increments $\Delta\Phi$ on the available Φs are exhausted. The problem to come next is to determine the statistical probability of N low-energy minima. These minima are first transformed to relative energies, taking the absolute minimum as zero energy conformation. Then we have $N - 1$ local minima L of energy E_L ($L = 1, N - 1$).

The existence probability p_L for state L is then given by the equation

$$p_L = w_L \exp[-E_L/kT]/\sum_L w_L \exp[-E_L/kT] \tag{28}$$

where w_L is a coefficient modeling the shape of the conformation minimum valley. It is evident that if two minima had the same numerical value, but the potential wells differed in shape, one being narrow and the other broad and shallow, the probability would be calculated to be the same if not for w_L. A criterion of evaluating w_L has been suggested in setting $w_L \sim |\det F''|$, where F'' is the above introduced Hessian matrix of second derivatives of the energy computed at the minima (where F'' is zero). This approximation shapes the energy hypersurface into a multidimensional paraboloid in the proximity of the minimum.

The "minimum region" is delimited by the sequence of all values of Φ adjacent to a hyperline joining all relative maxima encountered when moving out all possible directions from the center of the potential well. The procedure explores the region in question within the limits established by disregarding all conformations around the local minimum that are higher in energy (and therefore lower in probability) than a given upper threshold value. For example, this limit set at 4.0 kcal/mol corresponds to a Boltzmann factor $\exp(-\Delta E_L/kT)$ of 0.001 at 20°C.

With the obtained dimensions of the minimum well, the probability of existence $p(L)$ of a set of M_L conformational states around a minimum L of energy E_L among all N possible minimum states is given by the equation

$$p(M_L) = \sum_{}^{M_L} \exp[-E_m/kT] \ / \ \sum_{}^{M_L} \sum_{}^{N} \exp[-E_m/kT] \qquad (29)$$

where $m = 1, M_L$.

As an example, the outlines of the minimum wells of ethylmethylphosphate (EMP) are shown in Figure 16. Energies are expressed in kilocalories per mole on a scale whose zero corresponds to the global minimum conformation. The energies of the contour lines are given together with their probabilities.

Here we face a different way of dealing with MM results: the Boltzmann distribution function is an application of statistical mechanics to all allowed conformational microstates belonging to a rotational energy hypersurface and, thus, is immediately linked to the thermodynamic features of the molecular system.

Order-disorder conformational transitions in flexible structures can be modeled and predicted (strategic inference) at any temperature and correlated, for example, to phase transitions in materials and to the thermostability of molecules. In fact, one can ask how kinetic energy is dissipated over all available structural degrees of freedom of a molecule when heat is applied. Temperature-induced conformational transitions may be the principal factor in intramolecular energy distribution, thereby accounting for phase transitions in polymers, for example; or, on the contrary, they may play only a minor role in accepting excess energy in situations of hindered rotation, leaving the vibrational degrees of freedom as the only absorbing buffers. The strategic consequence for the researcher is a new aspect in the design of molecules having a programmed internal temperature "clock". Molecular structures can be predicted to be thermally unstable at specific temperatures because if only vibrational excitation is involved upon heating, at a certain critical temperature the weakest bond will break and thermolysis will occur in a predetermined manner.

The connection of the above distribution function to statistical thermodynamic quantities is well known to the reader and is summarized below.

Having p_{ij} as the probability of finding conformation i inside the cluster of all possible conformations within the energy well of the jth minimum, we can write the following equation for the Gibbs entropy (S_j):

$$S_j = -K \sum_i p_{ij} \ln p_{ij} \qquad (30)$$

For the internal energy U we have

$$U_j = \underline{E}_j = \sum_i p_{ij} E_i$$

$$= \sum_i n_{ij} \exp[-E_i/kT] E_i \ / \ \sum_i n_{ij} \exp[-E_i/kT] \qquad (31)$$

and for the heat capacity at constant volume C_V,

$$C_{Vj} = (\underline{E}_j/T)_V \qquad \text{and} \qquad C_{Vj}/T = (S_j/T)_V \qquad (32)$$

With the evaluation of these quantities a correct prediction of macroscopic phase transitions in polymers was achieved,[80] establishing in a novel way a correspondence between conformational transitions and structural changes inside the polymeric phase.

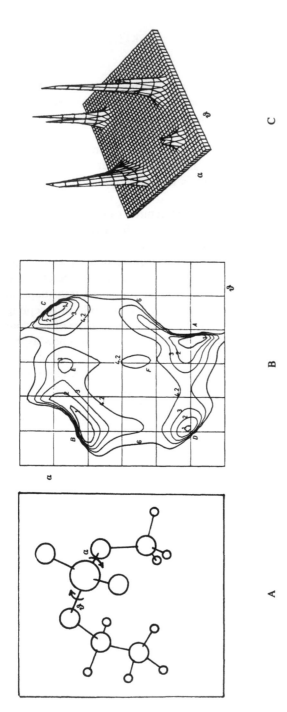

FIGURE 16. (A) The molecule EMP in the minimum energy conformation. (B) The corresponding map of the conformational energy is given as a function of the two dihedral angles α and ϑ. Energies are expressed in kilocalories per mole on a scale whose zero corresponds to the global minimum conformation. (C) Three-dimensional representation of the Botzmann distribution of probabilities of conformational states associated to the pair of rotations (α, ϑ) in EMP.

3. Molecular Dynamics

Another highly promising approach used to obtain conformational minima, especially in protein modeling and design, recently has found widespread acceptance: the molecular dynamics approach (MD).[81,82] It is used to calculate the dynamic possibilities of conformational changes and to improve our understanding of what structure really is. In our minds a chemical too often is envisioned to be the traditional image of an X-ray crystal structure, which seemingly informs us about the "real" shape of molecular species. On the contrary, the rigid appearance of a crystal structure is deceiving because it is the average virtual arrangement of a set of moving atoms in space. Atoms have their own velocities and move along certain trajectories. MD computations help in modeling these trajectories, which arise from the influence of intra- or/and intermolecular forces on the moving atoms. Thus, molecules undergo continuous fluctuations, minor or major changes in shape; enzymatic catalysis, for example, is believed only to occur due to the dynamic flexibility of the polypeptide chain, which is imagined to "wrap" dynamically around the substrate. Furthermore, the whole motion of a flexible molecular structure under the action of some external field can be studied by MD techniques as well.

The advent of supercomputers has attracted many researchers to attempt MD simulations of macromolecules in solution, a problem almost impossible to tackle with normal machines. In contrast to static MM, which aims at energy minimization, MD simulations yield changing conformations which are studied over given periods of time (called intervals) and at varying temperatures. The molecule is understood to be a dynamic, changing entity within the interval. The characteristic features of its changes reveal interesting novel descriptions of its behavior. As the concept of time immediately involves the concept of the future, MD, although probably the most numerically oriented methodology among those used in computer chemistry, resumes a computer chemical identity due to its strategic and informative capability coming from connecting events, i.e., from generating a whole tree of problem states around an initial state in the conformation space. The paths leading from node to node are correlated here to the different molecular morphologies existing in small time intervals along the global dynamic simulation. Here we are obviously dealing with a weighted graph, as some paths may be more favorable than others, meaning that some conformations will more likely be found at a certain temperature when coming from certain initial conditions. The mathematics of MD are similar to MM, with the addition of time t and velocity $\mathbf{v}$ to the energy equations.

The basic equation is Newton's second law of motion:

$$\mathbf{F} = \mathbf{ma} \tag{33}$$

where $\mathbf{a}$ is the acceleration vector, $\mathbf{F}$ is the force vector, which is the derivative of the potential E,

$$\mathbf{F}_i = -E(\mathbf{x}_i) = m_i d^2 \mathbf{s}_i / dt^2 \qquad (i = 1, n) \tag{34}$$

that acts on atom i of mass m_i, and $\mathbf{s}_i$ is the displacement vector of i along its trajectory. In order to obtain the new atomic positions $\mathbf{x}_i$, the above equations must be integrated simultaneously. We have the equation

$$d\mathbf{s}_i / dt = \mathbf{F}_i t / m_i + c$$

and at t = 0

$$d\mathbf{s}_i / dt = c = \mathbf{v}_{0i} \qquad \text{(initial velocity)}$$

and finally

$$ds_i \, / \, dt \; = \; \mathbf{v}_{0i} \, + \, \mathbf{a}_i t \tag{35}$$

With the trajectory increment we obtain after a time step Δt

$$\mathbf{x}_i \; = \; \mathbf{x}_{0i} \, + \, \mathbf{s}_i \; = \; \mathbf{x}_{0i} \, + \, [\mathbf{v}_{0i} \, + \, \mathbf{a}_i \Delta t] \Delta t \tag{36}$$

and for the new velocities

$$\mathbf{v}_i \; = \; \mathbf{v}_{0i} \, + \, \mathbf{a}_i \Delta t \tag{37}$$

The new values for position and velocity are inserted again in Equations 36 and 37, respectively, to compute the next piece of trajectory for each atom, and so on. After each iteration the accleration (i.e., the gradient of the applied potential) is recalculated by Equation 34 because it is itself a function of atomic coordinates. If in addition $E = f(t)$, then its actual value after each time interval also must be updated. In real calculations the time interval Δt is chosen in the range of picoseconds. The intramolecular potentials and forces are those described in Section II.B.2. If external forces are added (like inclusion of solvent shells, ions, or hydrogen bonding, for example), meaning that more realistic chemistry is added to the problem, other acceleration terms must be included accordingly.

The velocity of a particle is related to its kinetic energy, which in turn is related to temperature by kinetic theory:

$$n\mathrm{kT} \; = \; \sum_i \, \mathrm{m}_i \mathrm{v}_i^2 \tag{38}$$

Temperature is an essential feature in MD runs. If the MD simulation is started near 0K, then the initial velocity is approximately zero for all atoms, and the initial trajectory increment is calculated from the equations above. At the next time step, after recomputation of the new force magnitude at the incremented atomic positions, the new velocities and accelerations are used to obtain new shifts. When all time steps have been processed (or, more frequently, when your CPU time account has run out), the MD simulation comes to an end. The increase in velocity is equivalent to an increase in temperature. In practice, one wants to run simulations for situations either approximately at room temperature or, in special cases, at very high temperatures (to study the trajectories of major conformational transitions in proteins). At room temperature the atomic velocities differ significantly from zero; their magnitude is distributed in a Gaussian manner. Therefore, it could seem reasonable to start the MD calculation with initial velocities related to the desired temperature, but this approach leads to a highly unstable simulation, especially when large systems are processed — like proteins, which can contain up to 10,000 atoms (the hundreds of solvating water molecules not included). Instead, the following protocol has found acceptance and has proved to be more reliable. First, the molecular structure is energy minimized in the traditional way using static MM at 0K until only a small amount of internal potential energy is left (ca. 0.1 ± 0.01 kcal/mol). The molecular structure is said to be initially relaxed. The residual potential energy is then used to cause the atoms to start moving in the first stretch of time intervals of the MD run (around a few femtoseconds). Once this controlled motion has started, the temperature can be increased in small steps of about 10 to 20K, slowly reaching the desired final temperature. After each temperature increase, the Gaussian-distributed Maxwellian velocity increments (Equation 38) are attributed to the atoms randomly. With the new velocities a second stretch of MD simulation intervals is carried out, the temperature again increased, the additional velocity contributions randomly added to the atoms, and the MD run resumed. After the final temperature has been reached, the molecular dynamics run can be prolonged at will.

The outlined procedure is well suited to produce equilibrated intermediate structures, and it avoids pooling of excess kinetic energy in one region of the molecular system. The largest applications of MD are devoted to (1) simulation of protein dynamics and conformation changes, (2) protein solvation in aqueous environment, (3) ion transport through membranes, and (4) dynamics of liquids.* Concerning the conformational energy minimum of a protein, the following comment seems useful: there is no proof that the folding pattern of a natural enzyme is necessarily the one of lowest energy; it is just the one that best exploits its biocatalytic job. In addition, there is no realistic chance to obtain a complete map of all possible protein conformations, as millions of them exist with almost coalescing energy minima. This is the major reason why the initial enthusiasm[83] among researchers aiming to predict the 3-D structure of a protein from the linear amino acid sequence cooled down in the face of the enormous difficulties and crude assumptions made to model the system. Simplified representations of the polypeptide chain were introduced in an attempt to reduce the astronomical number of minima and to speed up energy calculations.

The method proved too weak to model reality (e.g., crystal structures of native globular proteins). Any search done by MM or MD cannot guarantee the reproduction of the X-ray conformation. It urges improvement of present techniques rather than development of new ones to obtain solid results in protein design. One obvious limitation is the available computing time: MD simulations involving up to 200 h of CPU time on fast machines are common. Even array processors in their current technological advancement are not yet enough to allow realistic complete simulations.** Also, a more detailed parameterization, with the same level of accuracy as we find in hydrocarbon parameterizations, must be accomplished to perform more refined calculations with better potentials (which provide for the necessary accelerations in MD). The energy function can be improved, explicitly incorporating all hydrogen atoms and all cross terms between bond lengths, bond angles, and torsion angles into the energy function. Polarization and solvent molecules (e.g., many hundreds of water molecules for a medium-sized protein) more realistically model the dielectric attenuation of the interaction between charged atoms. The novelty would be that a polarization term would change the energy equation from one where the independent contributions simply are summed up to one showing interrelated dependencies: the effective interaction between two atoms simultaneously depends on the locations and the features of all the surrounding atoms. Computation times would increase by a factor of 100 with these improvements, and this is within the reach of the supercomputer generation starting to appear on the market.

Help will also come from new conceptual advances in computer architecture, like parallel processing of different starting situations of a given system. With parallel supercomputers, for example, one might think of running MD simulations for a given system starting from different initial states. Every processor then explores only a specific region of the problem space, generating different solutions or quasisolutions which are evaluated by a central processor which, acting like an expert system, reshuffles duties and targets for the different processors until a satisfactory answer is found. Processor connection philosophies will have a fundamental impact on MD calculations in the future: for example, the hypercube concept pairwise and interactively connects processors. A hypercube of order 3 has 2^3 processors, which gives 8 units in total that can be imagined as being located on the corners of a cube. A hypercube of order 10 has 1024 processors. Extending this idea into a probably not so distant science fiction future, one could imagine the partitioning of our real physical space into subspaces, each covered by a single processor connected to a sufficient number of other

---

* For a review of points 1 and 2, see Reference 83; for points 3 and 4, see Reference 84, for example.
** Concerning the performance of array processors (and supercomputers in general), see Reference 85.

surrounding processors. Each processor then deals with the computation of only those quantities that affect the magnitude of the variables in its particular portion of physical space. We can imagine that if each atom in a protein is represented by a specific processor which is connected to a large number of other processors, it will "feel" the interactions exerted on it by the other "atoms" (which are processors). The breakthrough is that physical time and space can be modeled simultaneously. Such an exciting computer architecture requires a great deal of different programming work and a deep understanding of how data and procedures must be partitioned and reunited flexibly.

MD programs also can be used directly for more simple kinds of problems. One such example is given by the necessity of designing organic ligands for complexing metallic ions with molecular modeling and MD techniques.[50] An elastic force term, a_{EL}, is responsible for attraction of the user-selected donor atoms in the ligand structure toward the virtual docking zones in space around the metal ion, represented by the vertices of the coordination polyhedrons (derived from X-ray crystal data) characteristic of a given metal ion. The velocity of each atom moving toward its attributed coordination point is slowed down by a friction term (a_F) that simulates the energy loss required in the penetration of the solvation layers around the ion. The total acceleration a_t is then given as $a_t = a_{EL} - a_{FR}$. Plates 4A and B* show a nice example of the simulation of an Ni(II)-EDTA complex. The starting conformation of the complexing agent appears with randomized dihedral angles and is positioned about 15 Å from the nickel ion (Plate 4A), represented by the classical coordination octahedron. After completion of the MD simulation the EDTA structure wraps up the ion, very closely reproducing the geometry of the X-ray crystal structure.

MM and MD modeling are geared around an ensemble of different forces, some of which are directly related to atomic charges (Coulomb terms, dispersion terms). The description of a molecular structure cannot be complete without taking into account what is probably the most frequently used (and misused) parameter: the charge distribution.

C. ELECTRONIC MOLECULAR DESCRIPTORS

1. Introduction

A very important facet in the construction of a global molecular identikit consists of a family of electronic descriptors, like the atomic charge q, the effective atom-centered polarizability α, and the residual orbital electronegativity χ. They are fundamental parameters in the area of computer-assisted modeling of organic reactions, which is to be discussed later. Charges also play an important role in drug design and are used to calculate the dispersion and Coulombic forces in MM and MD simulations. Classical methods used to obtain molecular charge distributions are offered by the cornucopia of quantum chemical methods available at any degree of complexity in nearly every major chemical institute. Quantum mechanics (QM) approaches are, however, computational methods aimed primarily at assigning energies and geometries to molecular structures. They are characterized by very long computation times, a situation that greatly impedes their applicability when fast processing of a large number of molecules is required, such as (1) in synthesis design programs or (2) in the parameterization of a series of drugs in quantitative structure-activity relationship studies or in real time working with molecular modeling systems, where immediate responses are expected. Speed is certainly one central, irrevocable feature of chemical software in a preponderantly AI-oriented approach.

A number of non-QM empirical methods used to compute partial atomic charges have been proposed.[86-88] Among them there are some based on the well-known concept of electronegativity. One in particular has found acceptance in many molecular modeling systems because of its high processing speed and the quality of the predicted charges. It led to the

---

* Plates 4A and B follow page 62.

definition of a new parameter, the residual electronegativity, which was to prove of paramount importance in the computer-assisted modeling of chemical reactivity.

2. A Model for Sigma Charges

The concept of orbital electronegativity (OE)[89] is a natural extension of the traditional, atom-centered electronegativity. It is based on a theoretical definition of electronegativity χ (Equation 39). A combination of OE with a topological representation of a molecular structure (its constitutive graph) has proved successful in the construction of an empirical model for the computation of atomic partial charges in σ-bonded systems.[90-92] A later extension to π systems has been achieved as well.

$$\chi_{Ak} = \tfrac{1}{2}(I_{Ak} + E_{Ak}) \tag{39}$$

This equation states that the OE of an atomic orbital k of an atom A is given by the sum of its ionization potential, I_{Ak}, and its electron affinity, E_{Ak}. Each orbital in an atom within a particular hybridization state has its own specific OE value, χ_{Ak}.

The OE not only depends on the hybridization state of an atom, but also on its charge Q: an atom in its cationic state will show a higher electron attracting power than in an uncharged state, and an atom carrying a negative charge will conversely have a smaller OE than it would have in its neutral state. By inserting the appropriate values for I_{Ak} and E_{Ak} for a specific orbital k of a given atom A in its neutral, cationic, and anionic states into Equation 39, three OE values are calculated. The required I and E values are obtained from ground-state ionization potentials and valence-state promotion energies derived from spectroscopic data.[89,93]

The three points, $\chi_{Ak}(+)$, $\chi_{Ak}(0)$, and $\chi_{Ak}(-)$, do not lie on a straight line if plotted vs. atomic charge. There is nonlinear dependence of OE on Q. To model this dependence a second-order polynomial was chosen:

$$\chi_{Ak} = a_{Ak} + b_{Ak}Q_A + c_{Ak}Q_A^2 \tag{40}$$

The coefficients a_{Ak}, b_{Ak}, and c_{Ak} are characteristic parameters for each orbital of a specific atom in a particular hybridization state. They can be evaluated by solving the equation system

$$\chi_{Ak}(+) = a_{Ak} + b_{Ak} + c_{Ak} \qquad (Q = +1)$$

$$\chi_{Ak}(0) = a_{Ak} \qquad (Q = 0)$$

$$\chi_{Ak}(-) = a_{Ak} - b_{Ak} + c_{Ak} \qquad (Q = -1) \tag{41}$$

When two atoms interact and two orbitals overlap to form a σ bond, charge will flow from the more electropositive atom to the more electronegative one. This will increase the OE of the positively charged atom, according to Equation 40, and at the same time diminish the OE of the negatively charged bond partner. The OE difference between the two bonding orbitals therefore will decrease. It has been assumed first that charge separation occurs until total equalization of OE is obtained.[94-99] Total equalization means that all OEs of the atoms involved in forming the σ skeleton of a molecule reach a unique, final, common electronegativity value. In such a situation, as there are no more OE differences between linked atoms, the flow of charge would stop. Previous attempts to calculate partial atomic charges

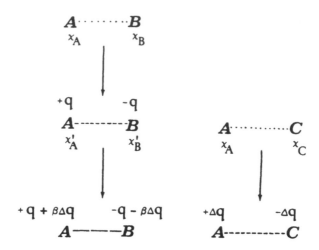

FIGURE 17.   A schematic representation of charge transfer between the orbitals of two atoms, A and B. After removal of negative charge from the more electropositive atom A, a decrease in overlap is generated that obstructs further charge flow. Comparison with a hypothetical system A-C shows the necessity of introducing a damping factor β to prevent orbital electronegativity from equalizing.

by any method leading to total equalization of *OE* yielded chemically unacceptable results.* Atoms of the same kind (e.g., all carbon atoms in a molecule, irrespective of their actual hybridization) would receive the same charge due to their identical final *OE* value. In ethanol, for example, the hydroxylic hydrogen and the methyl hydrogens would carry the same charge, in sharp contrast to all chemical and physical evidence. Furthermore, atoms of the same kind in isomeric molecules would also have the same charge (e.g., the carbon atoms in diethyl ether would show the same partial charge as those in butanol). In addition, isomeric groups would appear to have the same electronegativity at their attachment points. Their own specific chemical features, so important when discussing chemical reactivity and substituent effects, thereby become annihilated. Some attempts have been made to prevent total equalization of *OE*, but with limited success.

Partial equalization of *OE* is the key to one chemically acceptable model for the derivation of partial atomic charges. The following reasoning led to the creation of an iterative, empirical model resulting in the necessary *partial* equalization of *orbital electronegativity (PEOE)* within the molecular connection frame.

It should be realized that upon charge separation a change in overlap is generated along the σ bond. This density change is an obstacle to a further charge separation between two overlapping orbitals with different starting *OE* values. Figure 17 shows a simple model conception of bond formation which serves to illustrate this reasoning.

Consider a state in which two orbitals of atoms *A* and *B* are just beginning their interaction. The initial *OE* values χ_A and χ_B give rise to a certain charge separation q. The charge influences χ_A and χ_B to an extent quantified by Equation 40.

---

*   To justify the principle of total *OE* equalization, corrections to the isolated-atom electronegativity have been introduced. Due to the change in size and shape of an atom in a molecule, together with the Coulombic potential originating from the presence of neighboring charges, an effective electronegativity of an atom in a molecule is defined in some recent models. Being equalized to the effective electronegativity of all other atoms in the molecule, it contains enough information to obtain reliable charges. A theoretical basis for this revised principle of effective electronegativity equalization is given by Mortier et al.[112]

In the next step of this imaginary experiment the OE difference $|\chi_A' - \chi_B'|$ would cause a second, smaller charge separation Δq because obviously $|\chi_A' - \chi_B'| < |\chi_A - \chi_B|$. This process would go on until $|\chi_A' - \chi_B'| = 0$, i.e., until total equalization of OE. Let a second system consist of atoms A and C, with $\chi_C(0)$ such that the equation $\chi_A' - \chi_B' = \chi_A - \chi_C$ holds. At the initial stage of interaction within the A-C system no electric field is yet present and $q = 0$. However, in the A-B system, due to the partial charges $\pm q$, a decrease in overlap acts against the direction of charge flow. It will diminish the resulting overall effect of the electronegativity differences, the driving force for further charge separation. The result must be that $\Delta q_{AB} < \Delta q_{AC})$, or in an equivalent notation $\Delta q_{AB} = \beta \Delta q_{AC}$. The introduction of a damping factor β, empirically representing the progressive action of the decreasing overlap along bonds, is responsible for the avoidance of a total equalization of OE and of all its corollary shortcomings. The charge fraction q, transferred between two orbitals of atoms A and B during one particular iteration step s, is given by the equation

$$q^{<s>} = (^1/_2)^s \cdot (\chi_B^{<s>} - \chi_A^{<s>})/\chi_A^+ \tag{42}$$

where $(^1/_2)^s$ is the empirical damping factor and χ_A^+ is the OE of the specified orbital of atom A in its cationic state. In order to maintain a state of $Q_A = +1$, another atom attached to A must have an electronegativity which is at least as high as χ_A^+. As χ_A^+ relates an electronegativity value to the removal of one electron, it can be used to scale the difference in OE between two bond partners to a charge transfer quantified in electron units.

In the first iteration only half of the global electronegativity potential difference is allowed to exert its action. When the charge $q^{<1>}$ is generated, new electronegativities are calculated ($\chi_A^{<2>}$ and $\chi_B^{<2>}$) and Equation 42 reentered with the damping factor increased to $(0.5)^2 = 0.25$. This procedure is repeated until the charge transfer fades out, at the same time leaving the various OEs unequalized. For a diatomic molecule, Equation 42 suffices to compute the total charge Q of an atom. For larger molecules, all neighbors directly bonded to a central atom A must be taken into account simultaneously during each interaction step. For the orbitals m of neighbors M which are more electronegative than the orbitals k of atom A, whose charge is being computed, the value χ_{Ak}^+ must be taken; for less electronegative orbitals l of neighbors L, the constant χ_{Li}^+ must appear in the denominator. This leads to Equation 43 for the charge generated on atom A for each iteration step s:

$$q_A^{<s>} = \left[\sum_m (\chi_{Mm}^{<s>} - \chi_{Ak}^{<s>})/\chi_{Ak}^+ + \sum_l (\chi_{Li}^{<s>} - \chi_{Ak}^{<s>})/\chi_{Li}^+ \right](^1/_2)^s \tag{43}$$

After each step, the momentary total charge on A is calculated by the equation

$$Q_A^{<s>} = \sum q_A^{<s>} \tag{44}$$

The above equations model the influence of successive spheres of neighbors exerted on the actual atom A. The approach is purely topological in nature, and it can model only through-bond effects. Through-space effects would require a 3-D geometry of the molecule, but for most cases the inductive effects, which are so relevant in chemical reactivity reasoning, are reproduced very well by the $PEOE$ method. The final electronegativity for each atomic orbital when the charge flow vanishes is a very important parameter. It is called residual electronegativity (χ_R),[90,100] and it plays a central role in the modeling of chemical bond reactivity, as will be discussed in Chapter 7. Table 1 shows a complete calculation of propanol with the $PEOE$ method.

After six iterations the computation converges, the amount of charge shifted between connected atoms going below 1 millielectron. It should be repeated that the real advantage of this empirical model is the partial equalization of OEs. Therefore, isotopic atoms in

TABLE 1
A Complete Calculation of the Partial Atomic Charges for Propanol

Cycle	q_0	q_{C_1}	q_{C_2}	q_{C_3}	q_{H_1}	q_{H_2}	q_{H_3}	q_{H_4}	x_0	x_{C_1}	x_{C_2}	x_{C_3}	x_{H_1}	x_{H_2}	x_{H_3}	x_{H_4}
0									14.18	7.98	7.98	7.98	7.17	7.17	7.17	7.17
1	−338	122	−41	−61	20	20	20	175	9.97	9.13	7.61	7.43	7.30	7.30	7.30	8.24
2	−371	68	−31	−63	43	24	22	197	9.58	8.61	7.70	7.41	7.44	7.32	7.31	8.37
3	−385	53	−32	−63	51	27	23	204	9.42	8.47	7.69	7.41	7.48	7.33	7.31	8.42
4	−391	48	−32	−63	54	28	23	207	9.34	8.42	7.69	7.41	7.50	7.34	7.31	8.44
5	−394	45	−32	−63	55	28	23	209	9.31	8.40	7.68	7.41	7.51	7.35	7.31	8.45
6	−395	44	−33	−63	56	29	23	209	9.29	8.39	7.68	7.41	7.52	7.35	7.31	8.45

Note: After six iterations, the calculations converge to final electronegativity values which are different for all topologically nonequivalent atoms.

different environments become distinguishable. In the propanol example it is evident that all constitutionally nonequivalent atoms receive different charges as well as different residual electronegativities. The attenuation of the inductive effect of the hydroxylic group over an increasing number of bonds is visible both on the carbon atoms and on the hydrogens.

Computations are extremely fast: a few tenths of a second for a medium-sized organic molecule are normal CPU times on mainframe computers. It is noticeable that, in contrast to QM methods, within the *PEOE* calculational approach the computation times increase linearly with the number of atoms n in a molecule (an increase with n^2 up to n^4 is the rule in QM approaches).

3. The Model for Pi Electrons

The π electron system is at least as important as the σ skeleton for the reactivity of a molecule. It therefore calls for a natural extension of the *PEOE* approach to π orbitals using π-orbital electronegativities (*POE*). The values for p_z orbitals and free electron pairs are available in the literature.*

Initially, however, it appeared quite strange that no model had ever been presented for the computation of π charges by means of *POE*, while so many attempts had been made for σ electrons. The reason was found immediately when an attempt was made to generate a charge transfer between a free electron pair of a generic donor atom and a conjugated π system using the available ground-state π-orbital electronegativities. Let us consider the π bond scheme in a generic vinyl ether molecule consisting of a C=C next to the free electron pair of an oxygen atom.

$$O\!-\!C\!=\!C$$

From very different experiments (bond lengths, nuclear magnetic resonance [NMR], electron spectroscopy for chemical analysis [ESCA]) and from theory it has been established that an interaction between the free electron pair of the heteroatom and the adjacent π system occurs. Resonance theory even uses extreme notations like $O^+\!=\!C\!-\!C^-$ to symbolize the electron donating capability of the oxygen atom (or other atoms having comparable donor properties). It therefore could be reasonably expected that the π charge shift might be modeled by the two *POE*s of the $p_z(C_1)$ orbital and of the oxygen free electron pair. This, however, turns out to be impossible. The *POE* of a free electron pair of oxygen is larger than the carbon p_z *POE* (7.91 and 5.6 eV, respectively). In other words, in the initial state [$q(\pi) = 0$] the free electron pair of oxygen is more electronegative than the carbon orbital, and consequently a charge flow in the direction $O \to C$ cannot be predicted in such a simple approach. (From the mere difference of the two POEs one should even predict a reverse charge flow, $C \to O$!)

What was just said about oxygen is valid for all other common donor atoms (e.g., nitrogen, sulfur, and the halogens). Also, no correct π donation toward electron-accepting π systems can be predicted for them.

A closer scrutiny of the problem led out of this impasse. An artificial separation of the σ and π systems was postulated tentatively, similar to what is often encountered in some semiempirical QM methods dealing with π electrons.

The π electrons, being bonded less tightly to the atomic core than the σ electrons, are more polarizable. They feel the changes in nuclear screening arising from the changing magnitude of the more internal σ charge. As a consequence, a change in the effective values of the ionization potential and electron affinity for π electrons and free electron pairs takes place. Thus, a positively charged atom will have its π electrons more strongly attracted to

---

* See Reference 61 for neutral-state orbital electronegativities.

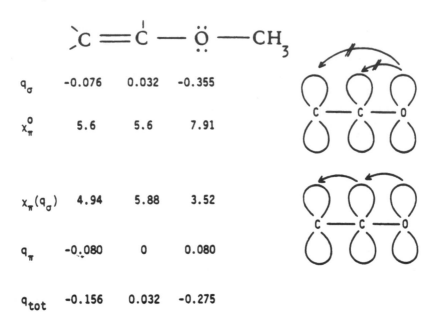

q_σ -0.076   0.032   -0.355

χ_π^0 5.6   5.6   7.91

$\chi_\pi(q_\sigma)$ 4.94   5.88   3.52

q_π -0.080   0   0.080

q_{tot} -0.156   0.032   -0.275

FIGURE 18.   Using neutral-state *POE* values no donor effect can be predicted in π systems. Only after introducing a dependence from the σ charge for the π-orbital electronegativity does a mesomeric donor effect become feasible.

the core than in its neutral state, leading to increased I and E values. On the other hand, a negatively charged atom will exert a kind of repulsion on its π electron(s), leading to lower I and E values. The *POE* therefore was described as a function of the σ charge of the corresponding atom. In calculating the *POE* of p_z electrons and of free electron pairs under the influence of unit charges in the σ level ($q_\sigma = 0, \pm 1$), three *POE* values are obtained using the expression given in Equation 40. These values are fitted again by a simple second-order polynomial:

$$\chi(\pi)_i = a(\pi)_i + b(\pi)_i q(\sigma)_i + c(\pi)_i q(\sigma)_i^2 \tag{45}$$

This equation states that the π-orbital electronegativity of atom i depends on the σ charge of this atom. The model is called sigma-dependent pi orbital electronegativity (*SD-POE*).[101] It indicates the chronology of charge calculations: the σ charges must be evaluated first, and then they are used to determine the starting values for the various $\chi(\pi)$ for the subsequent π charge calculation.

A dramatic reversal of the role of $\chi(\pi)$ originates from this simple conceptual revision. Its application in the previous vinyl ether example proves its validity. The σ charge of the oxygen atom (see Figure 18) is -355 millielectrons, and its neutral state *POE* $\chi(\pi^0)$ is 7.91 eV. After recomputing according to Equation 45, a value of only 4.94 eV is obtained. The σ charge of the adjacent carbon atom is $+32$ millielectrons, which changes its primitive $\chi(\pi^0)$ value from 5.6 to 5.88 eV. The carbon orbital is now more electronegative than the oxygen free electron pair: a $+M$ donor effect now becomes feasible and is quantified by the equation

$$q(\pi) = [\chi(\pi)_j - \chi(\pi)_k]/\chi(\pi^+)_k \tag{46}$$

This equation does not have an iterative structure. The π charge between two interacting atoms j and k is computed in one step. This one-step approach recently was extended to an iterative model in which the global charge ($\sigma + \pi$) was used to readjust the charge pattern in the σ skeleton, which in turn would cause a minor charge shift in the π sphere.[102] The differences between the two models are purely numerical and not conceptual, and the subsequent model refinement will not be discussed further here.

When working on a topological representation of a molecule, on which simple valence-bond mesomeric structures are constructed, another question must be solved empirically: on which atom(s) in the π system should the generated charge be localized? The mesomeric valence-bond structure of vinyl ether indicates a 100% charge allocation to the C_2 carbon atom, a scheme followed in many of the computer programs reproducing this calculational model. Certainly a total neglect of C_1 may not correspond to reality, but further calculations have shown that an eventual internal rearrangement of the π charge can be accomplished by an internal polarization of that charge between the two carbon atoms, as they have different *POE*s. These differences are, however, generally small and do not change the overall aspect of the π charge distribution pattern significantly. Also, some spectroscopic evidence has been invoked to sustain the negligible presence of donor π charge on C_1.

a. Delocalized Systems

So far an introduction into the calculation of π charges in isolated π systems was given. The extension to larger conjugated or delocalized systems takes into account the specific features of the processed π systems. A short presentation of the general model for extended π systems follows.

When a certain $\pm M$ substituent **X** is conjugating with an adjacent π bond

$$X\text{–}C_1\text{=}C_2\text{–}C_3\text{=}C_4\text{–} \ldots C_{n-1}\text{=}C_n \qquad (X = +M, -M \text{ group})$$

$$L = 1 \quad L = 2 \qquad L = n/2 = R$$

(or a chain of π bonds), the orbitals considered for entering Equation 46 are always the topologically nearest interacting orbitals between **X** and the connected π system. If **X** is a $+M$ donor atom with a free electron pair, the interacting orbitals are the electron pair and the $p_z(C_1)$ orbital. If **X** is a $-M$ group (C=O, C=N, NO_2, etc.), the interacting orbitals are the p_z orbital of the atom in the $-M$ group bonded to C_1 and, again, the $p_z(C_1)$ orbital. In line with the symmetry of the highest occupied MO (HOMO) and in accordance with the valence bond theory, the charge generated between the adjacent orbitals j and k is transmitted to C_2, C_4,... and down to the last resonating center, C_n. In the case of such conjugated systems, one must be aware of the presence of single bonds between the π subsystems. The single bonds are periodic potential barriers that prevent complete and "frictionless" transmission of the generated π charge from **X** to C_n. The more single bonds situated between the π charge generator **X** and a resonance center r, the smaller the fraction of π charge attributed to r. A QM description of the decreasing amount of π charge transferred along polyene systems[103] introduces a transmission coefficient ϕ^L, where L is automatically the number of single bonds between r and **X**. It has been shown that the influence of a $\pm M$ group on a system of R conjugated π bonds follows a geometric decrease like

$$q(\pi)^{(1)} : q(\pi)^{(2)} : q(\pi)^{(3)} : \ldots q(\pi)^{(R)} = \phi^1 : \phi^2 : \phi^3 : \ldots \phi^R$$

The total charge $q(\pi)$ generated between j and k is partitioned over all R resonating atoms following a normalized attenuation coefficient ϕ^L. Thus, we have for the computed charge at atom r

$$q(\pi)_r = N_R \phi^L [\chi(\pi)_j - \chi(\pi)_k]/\chi(\pi^+)_k$$

$$\left(N_R = 1/\sum_L^R \phi^L \right) \tag{47}$$

In aromatic systems there is complete delocalization of π electrons. Once the π interaction gives rise to a certain charge $q(\pi)$, this charge has to be split over the resonating ortho and para positions. The use of an attenuation factor ϕ^L and a distance L from the generating $\pm M$ group becomes meaningless. Due to the different σ charges in the ortho and para positions (with respect to **X**), the local $\chi(\pi)$ will have different values for the respective p_z orbitals. In the case of $+M$ groups bonded to an aromatic system, the higher the $\chi(p_z)$ value, the higher the probability of finding the π charge at this specific position. A high $\chi(p_z)$ is more likely to stabilize the negative excess π charge donated from the $+M$ group. On the contrary, for $-M$ groups, which are π electron attracting, the lower the $\chi(p_z)$ value, the more likely the positive π charge will be found on a certain atom of the aromatic system. The π charge fraction computed in the *SD-POE* model for a generic aromatic resonance center i is proportional to a statistical factor W and is given by the following equations:

$$+M: \quad q(\pi)_i = W^+(i)q(\pi)$$

$$-M: \quad q(\pi)_i = W^-(i)q(\pi) \tag{48}$$

where the weighting factors are given by

$$W^+(i) = \chi(\pi)_i / \sum_r^R \chi(\pi)_r \quad \text{and} \quad W^-(i) = [\chi(\pi)_i]^{-1} / \sum_r^R [\chi(\pi)_r]^{-1}$$

The sums go over all R resonating atoms r. The absolute total sum of the fractional π charges at the centers i equals the π charge at the substituent **X**.

Algorithmic Digression

The previously proposed empirical model for π charge calculation relies on the topological representation of a molecular structure. Thus, the previously explained internal representation for a 2-D structure can be used directly to construct the necessary algorithmic steps to trace the π frame and execute the calculations leading to the final charges.

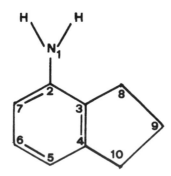

A general π charge program therefore must contain the following (very simplified) steps:

1. Determine all atoms with a multiple bond (i.e., if *BM* (3,*i*) > 1, set a flag on these atoms: MULT_BOND(*i*) = TRUE).
2. Determine all generators of +M and −M effects and set a flag PLUS_M or MIN_M on them (example: nitrogen in the aniline derivative).

IF AT_NUMBER(*i*) = 7 AND FREE-ELECTRON(*i*) = 2
  AND MULT_BOND(*i*) = FALSE
    THEN PLUS_M(*i*) = TRUE

3. Use a recursive procedure to determine the graph of all conjugated multiply bonded atoms starting from the PLUS_M (or MIN_M) atoms. Set flag CONJ_MULTB on atoms that are found.

4. Using the vector RING marking ring atoms, find the aromatic ring atoms. Set flag ARO on these atoms.
$$RING = (0,1,1,1,1,1,1,1,1,1)$$
$$CONJ\_MULTB = (0,1,1,1,1,1,1,0,0,0)$$
$$ARO = RING\ AND\ CONJ\_MULTB = (0,1,1,1,1,1,1,0,0,0)$$

5. Find alternating ARO atoms with respect to the actual ±M group and store these atoms in vector RES_AT.

$$RES\_AT = (3,5,7)$$

6. Use $\chi(\pi)$ for the interacting PLUS_M (or MIN_M) and REST_AT centers to compute the π charges.

This short digression underlines once more the practicality of Boolean operations for the manipulation of chemical structures in their topological representation. They are very fast operations inside a computer and, if intelligently geared, can act as a powerful selective filter for extracting very specific features from the global molecular structure information.

4. Correlations with Experimental Quantities

To establish the performance of these approaches based on *OE*, exhaustive testing was done in correlating the computed charges with experimental quantities known to be related to partial atomic charges, like C-1s ESCA chemical shifts,[92] ^{1}H-NMR chemical shifts,[104] dipole moments,[105] NMR coupling constants,[106] and some reactivity parameters.[107]

All investigations showed the reliability of the computed partial atomic charges, which in some cases proved to be more accurate in predicting experimental parameters than the more time-consuming QM methods.[92]

In all of the studies given in the references the role of partial atomic charges in explaining physicochemical quantities has been discussed extensively. However, it is wrong to try to intuitively correlate charges with atomic or molecular properties just because we are in the habit of doing it, neglecting other important factors that might interfere. Charges, as calculated in the presented models, are ground-state descriptors. They can only be used to model other parameters dependent on the molecular ground-state charge distribution. Whenever additional factors are involved, like excited states and magnetic anisotropies in NMR or polarizability and charge stabilization in chemical reactivity, the sharply defined role of atomic charges becomes blurred. This apparently trivial statement has its importance in the prediction of chemical reactivity if done by a computer, which, lacking intuition, must rely

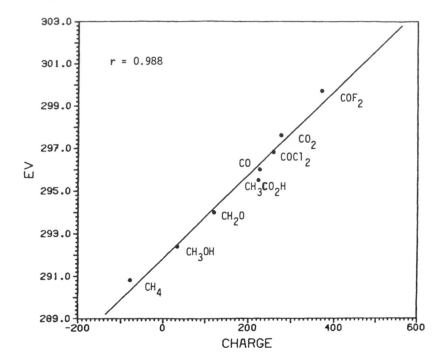

FIGURE 19.   A plot of the correlation between the charges of a series of increasingly oxidized carbon atoms and their corresponding C-1s ESCA shifts.

on the right numbers to draw conclusions. Especially in the area of reactivity of bonds, the bare atomic charge value taken as such proved to be a weak descriptor.

The following correlations are examples provided to illustrate the usefulness of computed *OE* charges in those cases where charge is the major factor influencing a specific physicochemical feature of the measured system. Figure 19 shows a plot of the charges of carbon atoms in a series of organic compounds vs. their corresponding ESCA C-1s energies. The scope of the study is to investigate more quantitatively the concept of oxidation of a carbon atom. This term is often used when discussing organic reactions, especially some nucleophilic reactions occuring at a carbonyl group. How positive really is a carbon atom with increasing substitution by electronegative partners? The plot shows that a linear relationship exists between computed charges and their experimental sensor, the C-1s electron binding energy, from methane to the most oxidized carbon atom in COF_2. This result is to be understood as (1) a means for predicting ESCA shifts for unmeasured structures and (2) a calibration test for the "goodness" of the *OE* charge models. This investigation relates a calculated ground-state property with a quasi-ground-state property, namely, the binding energy of an inner electron. This assumption is only true within the "frozen orbital" approximation,[108] in which the electronic relaxation of neighboring electrons is regarded as negligible. In this approach the binding energy of a 1s electron, E_{1s}, is linearly dependent on the partial atomic charge of the measured atom:

$$E_{1s} = kq + E^0 \qquad (49)$$

where E^0 is the energy of the 1s electron in an atom with zero excess charge. It should be pointed out that the ranking of the carbon atoms according to their degree of oxidation does not have to follow their actual reactivity tendency, which is dominated by other contributions as well.

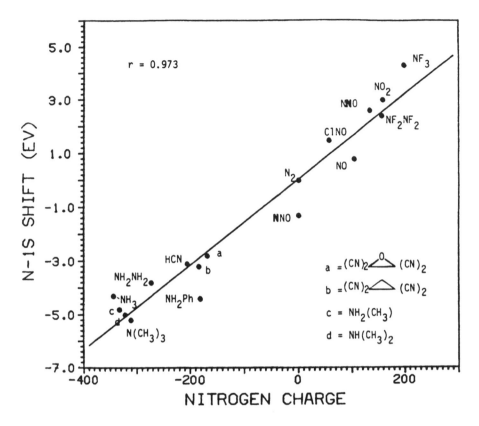

FIGURE 20. Correlation of atomic charges at the nitrogen atoms with N-1s ESCA energies.

Figure 20 shows the correlation of nitrogen charges with their N-1s ESCA shifts. The trend of increasing binding energy of the core electrons along with an increasing positive charge is reproduced very well by the model. In contrast to a carbon atom, nitrogen has a free electron pair which undergoes polarization quite easily. It becomes evident when studying the worst point in the correlation, aniline, that delocalization and subsequent relaxation of the π molecular orbitals influence the magnitude of the N-1s binding energy of an unperturbed 1s electron. This is probably the border situation for the validity of Koopman's theorem, which requires localized "frozen" orbitals for a linear dependence of energy on charge in the ionization experiment.

The mutual position of the four points representing NH_3, NH_2CH_3, $NH(CH_3)_2$, and $N(CH_3)_3$ is orthogonal to the main regression line. This means that in the ground state a methyl group has less electron donating power than a hydrogen atom, but becomes a better electron donator when stabilization of a vicinal positive charge becomes necessary. An alkyl group can be seen as an inductive reservoir of negative charge. For this reason the binding energy of the N-1s electron in ammonia appears to be higher than in trimethylamine.

The plot in Figure 20 very simply demonstrates the attenuation of the inductive effect along a chain of σ bonds. The charges of hydrogen atoms in a variety of halogenated organic compounds are correlated with their ^{1}H-NMR chemical shifts. The decrease in diamagnetic shielding of the protons follows their increasing partial charges and reflects the through-bond inductive effect. As long as other contributions are negligible (responsible for inclusion of excited molecular states), the chemical shifts are good sensors for atomic charges. The empirical OE-based model clearly reproduces the inductive effect exerted by a substituent on a chain of σ bonds, as can be seen, for example, from the hydrogen atoms of chloropropane

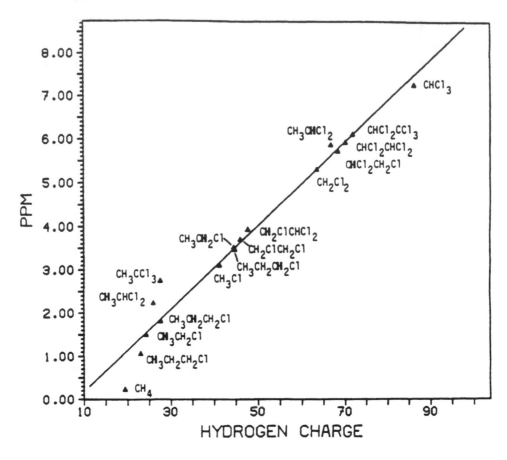

FIGURE 21.   A plot of hydrogen charges vs. their ^{1}H-NMR chemical shifts for a series of halogenated compounds. The increasing charge at the hydrogen atom well reproduces the linear increase in downfield shift.

in positions 1, 2, and 3 to the halogen, which are all close to the regression line. This is shown in Figure 21.

5. Effective Polarizability

Another useful parameter for adding a stone to the molecular identikit puzzle through fast topological computational models is the polarizability α. It is an important parameter in computer modeling of chemical reactivity. The estimation of the mean molecular polarizability α is given by the equation

$$\alpha = 4\left(\sum_i \tau_i\right)^2 / N \qquad (50)$$

where N is the total number of electrons in the molecule and τ_i is the polarizability increment for each atom i, characteristic for each atom type and its hybridization state.

The mean molecular polarizability can also be calculated by the Lorenz-Lorentz equation from the refractive index (n_D), the molecular weight, and the density of a compound. This is a demonstration that τ_i can be derived from elementary molecular properties.

Polarizability is a measure of how easily a distortion of a polar system occurs upon electromagnetic perturbation. The stabilization energy due to the interaction between an external charge and the induced dipole is highly distance dependent. The equations of classical electrostatics allow the computation of this energy. However, if the inducing charge resides inside the molecular frame, the situation is less clear. To model the stabilization caused by an internal polarization (a situation encountered, for example, when modeling the heterolytic rupture of chemical bonds), the above equation was modified,[109] yielding

$$\alpha_d = 4\left(\sum_i d^{n-1}\tau_i\right)^2/N \tag{51}$$

where a damping factor d^{n-1} models the attenuation of the stabilizing effect along n bonds away from the charged polarizing atom i ($0 < d < 1$). Equation 51 defines an effective polarizability (α_d). The damping factor is responsible for different α_d values for the same molecule, depending on where the charge center is located (e.g., where a specific bond breaks). Here we have an extension of a simple zero-order additivity scheme (Equation 50) to a first-order scheme (Equation 51), as in the case of the model for charge calculation. A zero-order additivity scheme yields a global molecular property by addition of atomic properties, whereas a first-order additivity scheme also involves bond properties. A second-order scheme would involve group properties. It is notable that the zero-order approximation delivers one single global value for α, regardless of the counting sequence of the atoms. Conversely, the first-order approach takes the bonds surrounding the charged atom into account and provides a local, site-specific property. The role of effective polarizability, taken alone or in a linear function with q and χ_R, has been analyzed from the perspective of chemical reactivity, and its usefulness in modeling basic organic reaction steps has been established.[110] One of the central reaction classes in organic chemistry, proton abstraction and proton acquisition, quantified by the pK_A and proton affinity (PA) values, has been described efficiently by functions of the type

$$PA, pK_A = F(\chi_R, \alpha_d)$$

These types of functions, with some extensions to other descriptors, are key variables in one particular successful approach to computer-assisted prediction of organic reactions, a subject to be discussed later. The effective polarizability introduced here provides a quantitative explanation[110,111] for the puzzling result that gas-phase protonation of alcohols (positive charge development) and gas-phase acidity of alcohols (negative charge development) are both favored in the order methanol < ethanol < isopropyl alcohol < t-butanol. The polarizability effect results in the stabilization of both positive and negative charges. In solution, stabilization due to polarizability is mixed with the solvent-provided contribution that adds other charge-stabilizing mechanisms.

REFERENCES

1. **Keverling Buisman, J. A., Ed.**, *Biological Activity and Chemical Structure*, Elsevier, Amsterdam, 1977.
2. **Keverling Buisman, J. A., Ed.**, *Strategy in Drug-Research*, Elsevier, Amsterdam, 1982.
3. **Olson, E. C. and Christoffersen, R. E., Eds.**, *Computer-Assisted Drug Design*, ACS Symp. Ser., Vol. 205, American Chemical Society, Washington, D.C., 1979.
4. **Ariens, E. J., Ed.**, *Drug Design*, Academic Press, New York, 1980.
5. **Jolles, G. and Woolridge, K. R. H., Eds.**, *Drug Design: Fact or Fantasy?*, Academic Press, London, 1984.
6. **Hansch, C.**, On the structure of medicinal chemistry, *J. Med. Chem.*, 19, 1, 1976.
7. **Martin, Y. C.**, A practitioner's perspective of the role of quantitative structure-activity analysis in medicinal chemistry, *J. Med. Chem.*, 24, 229, 1981.
8. **Topliss, J. G., Ed.**, *Quantitative Structure-Activity Relationships of Drugs*, Academic Press, New York, 1983.

9. **Randic, M.,** On the recognition of identical graphs representing molecular topology, *J. Chem. Phys.,* 60, 3920, 1974.

10. **Balaban, A. T. and Harary, F.,** The characteristic polynominal does not uniquely determine the topology of a molecule, *J. Chem. Doc.,* 11, 258, 1971.

11. **Mackay, A. L.,** On rearranging the connectivity matrix of a graph, *J. Chem. Phys.,* 62, 308, 1975.

12. **Herndon, W. C.,** The characteristic polynomial does not uniquely determine molecular topology, *J. Chem. Doc.,* 14, 150, 1974.

13. **Smith, G. E., Ed.,** *The Wisswesser Line-Formula Chemical Notation,* McGraw-Hill, New York, 1968.

14. **Johns, T. M. and Clare, M.,** Wisswesser line notation as a structural summary medium, *J. Chem. Inf. Comput. Sci.,* 22, 109, 1982.

15. **Gasteiger, J.,** EROS User Manual, Version 3.2, Technical University Munich, Munich, Federal Republic of Germany.

16. **Hippe, Z.,** Manipulation of chemical structures within a computer, in *Data Processing in Chemistry,* Studies in Physical Theoretical Chemistry, Vol. 16, Hippe, Z., Ed., Elsevier, Amsterdam, 1981, 249.

17. **Hippe, Z., Achmactowicz, O., Jr., and Hippe, R.,** Some problems of computer-assisted discovery of organic syntheses, in *Data Processing in Chemistry,* Studies in Physical and Theoretical Chemistry, Vol. 16, Hippe Z., Ed., Elsevier, Amsterdam, 1981, 207.

18. **Frerejaque, M.,** Condensation d'une molecule organique, *Bull. Soc. Chim. Fr.,* p. 1008, 1939.

19. **Plotkin, M.,** Mathematical basis of ring-finding algorithms in CIDS, *J. Chem. Inf. Comput. Sci.,* 11, 60, 1971.

20. **Zamora, A.,** An algorithm for finding the smallest set of smallest rings, *J. Chem. Inf. Comput. Sci.,* 16, 40, 1976.

21. **Fugmann, R., Doelling, U., and Nickelsen, H.,** Das Problem chemischer Ringstrukturen aus topologischer Sicht, *Angew. Chem.,* 79, 802, 1967.

22. **Corey, E. J. and Petersson, G. A.,** An algorithm for machine perception of synthetically significant rings in complex cyclic organic structures, *J. Am. Chem. Soc.,* 94, 460, 1972.

23. **Bersohn, M.,** An algorithm for finding the synthetically important rings of a molecule, *J. Chem. Soc. Perkin Trans. 1,* 12, 1239, 1973.

24. **Esack, A.,** A procedure for rapid recognition of the rings of a molecule, *J. Chem. Soc. Perkin Trans. 1,* 12, 1120, 1975.

25. **Wipke, W. T. and Dyott, T. M.,** Use of ring assemblies in a ring perception algorithm, *J. Chem. Inf. Comput. Sci.,* 15, 140, 1975.

26. **Gasteiger, J. and Jochum, C.,** An algorithm for the perception of synthetically important rings, *J. Chem. Inf. Comput. Sci.,* 19, 43, 1979.

27. **Schubert, W. and Ugi, I.,** Constitutional symmetry and unique descriptors of molecules, *J. Am. Chem. Soc.,* 100, 37, 1978.

28. **Jochum, C.,** Algorithmen zur Auswertung konstitutioneller Information organisch-chemischer Strukturen, Ph.D. thesis, Technical University Munich, Munich, Federal Republic of Germany, 1978.

29. **Jochum, C. and Gasteiger, J.,** Canonical numbering and constitutional symmetry, *J. Chem. Inf. Comput. Sci.,* 17, 113, 1977.

30. **Dittmar, P. G., Mockus, J., and Couvreur, K. M.,** An algorithmic computer graphics program for generating chemical structure diagrams, *J. Chem. Inf. Comput. Sci.,* 17, 186, 1977.

31. **Carhart, R. E.,** A model-based approach to the teletype printing of chemical structures, *J. Chem. Inf. Comput. Sci.,* 16, 82, 1976.

32. **Shelley, C. A.,** Heuristic approach for displaying chemical structures, *J. Chem. Inf. Comput. Sci.,* 23, 61, 1983.

33. **Nordlander, J. E., Bond, A. F., IV, and Bader, M.,** ATCOOR: a program for calculation and utilization of molecular atomic coordinates from bond parameters, *Comput. Chem.,* 1, 209, 1985.

34. **Gund, P.,** Present and future computer aids to drug design, in *X-Ray Crystallography and Drug Action,* Horn, A. S. and DeRanter, C.J., Eds., Clarendon Press, Oxford, 1984, 495.

35. **Potenzone, R. P., Jr., Cavicchi, E., Weintraub, H. J. R., and Hopfinger, A. J.,** Molecular mechanics and the CAMSEQ processor, *Comput. Chem.,* 1, 187, 1977.

36. **van der Lieth, C. W., Carter, R. E., Dolata, D. P., and Liljefors, T.,** RING — a general program to build ring systems, *J. Mol. Graphics,* 2(4), 117, 1984.

37. **Gann, L. and Gasteiger, J.,** Eine Verarbeitung der R,S und E,Z Nomenklatur zur Spezifikation der Sterochemie, in *Software-Entwicklung in der Chemie,* Gasteiger, J., Ed., Springer-Verlag, Berlin, 1987, 17.

38. **Cahn, R. S., Ingold, C., and Prelog, V.,** Spezifikation der molekularen Chiralitat, *Angew. Chem.,* 78, 413, 1966.

39. **Prelog, V. and Helmchen, G.,** Bases of the CIP system and proposal for a revision, *Angew. Chem.,* 94, 614, 1982.

40. **Motherwell, S.**, PLUTO, a Program for Plotting Molecular and Crystal Structures, Chemical Laboratory, Cambridge University, Cambridge, U.K.

41. **Porter, T.**, Spherical shading, *Comput. Graphics*, 12, 282, 1978.

42. **Honegger, P. L.**, Generation of molecular surfaces, *J. Mol. Graphics*, 1(1), 9, 1983.

43. **Max, N.**, Computer representation of molecular surfaces, *J. Mol. Graphics*, 2(1), 8, 1984.

44. **Morffew, A. J.**, Bibliography for molecular graphics, *J. Mol. Graphics*, 1(1), 17, 1983.

45. **Hiller, C. and Gasteiger, J.**, Ein Automatisierter Molekülebaukasten, in *Software-Entwicklung in der Chemie*, Gasteiger, J., Ed., Springer-Verlag, Berlin, 1987, 53.

46. **Marsili, M.**, Technical Report N. 88/1, Tecnofarmaci Computer Chemistry Project, Tecnofarmaci SpA, Pomezia, Italy, 1988.

47. **Cory, M., Jr.**, MATCHMOL, an interactive computer graphics procedure for superposition of molecular models, *J. Mol. Graphics*, 2(2), 39, 1984.

48. **Taylor, G.**, FITZ — interactive graphics tool for investigating molecular symmetry and homology, *J. Mol. Graphics*, 1(1), 5, 1983.

49. **Tucker, J.**, Designing molecules by computer, *High Technol.*, January, 52, 1983.

50. **Marsili, M., Marengo, E., Salomone, M., del Buono, C., Cammarata, F., Scavia, G., and Caglioti, L.**, Computer chemistry at Tecnofarmaci, *Chim. Ind. (Milan)*, 5, 25, 1987.

51. **Marsili, M., Floersheim, P., and Dreiding, A. S.**, Generation and comparison of space-filling molecular models, *Comput. Chem.*, 7, 175, 1983.

52. **Marsili, M. and Floersheim, P.**, *DRACO: An Interactive System for Boolean Representation and Manipulation of 3D Molecular Models*, Anal. Chem. Symp. Ser., Vol. 15, Elsevier, Amsterdam, 1983, 332.

53. **Gund, P., Andose, J. D., Rhodes, J. B., and Smith, M. G.**, Three-dimensional molecular modeling and drug design, *Science*, 208, 1425, 1980.

54. **Dyott, T. M., Stuper, A. J., and Zander, G. S.**, MOLY — an interactive system for molecular analysis, *J. Chem. Inf. Comput. Sci.*, 20, 28, 1980.

55. **Gund, P.**, Three-dimensional pharmacophoric pattern searching, *Prog. Mol. Subcell. Biol.*, 5, 117, 1977.

56. **Marshall, G.**, Drug design, computer graphics and receptor modeling, *Pharmacochem. Libr.*, 6, 129, 1983.

57. **Langridge, R., Ferrin, T. E., Kuntz, I. D., and Connolly, M. L.**, Real-time color graphics in studies of molecular interactions, *Science*, 211, 661, 1981.

58. **Feldmann, R. J., Bing, D. H., Furie, B. C., and Furie, B.**, Interactive computer surface graphics approach to study the active site of bovine trypsin, *Proc. Natl. Acad. Sci. U.S.A.*, 75, 5409, 1978.

59. **Barino, L. and Scordamaglia, R.**, Characterization and predictive evaluation of active compounds through molecular fit techniques, *Chim. Ind. (Milan)*, 11, 118, 1986.

60. **Tosi, C., Scordamaglia, R., Barino, L., Ranghino, G., Fusco, R., and Caccianotti, L.**, Development and application of computational chemistry techniques, *Chim. Ind. (Milan)*, 1, 1, 1987.

61. **Horvath, C., Melander, W., and Molnar, I.**, Solvophobic interactions in liquid chromatography with nonpolar stationary phases, *J. Chromatogr.*, 125, 129, 1976.

62. **Sinanoglu, O.**, The solvophobic theory for the prediction of molecular conformations and biopolymer bindings in solutions with recent direct experimental tests, *Int. J. Quantum Chem.*, 18, 391, 1980.

63. **Humblet, C. and Marshall, G. R.**, Three-dimensional computer modeling as an aid to drug design, *Drug Dev. Res.*, 1, 409, 1981.

64. **Verloop, A., Hoogenstraten, W., and Tipker, J.**, Development and application of new steric substituent parameters in drug design, in *Drug Design*, Vol. 7, Ariens, E. J., Ed., Academic Press, New York, 1976, 165.

65. **Hopfinger, J.**, A QSAR investigation of dihydrofolate reductase inhibition by Baker triazines based upon molecular shape analysis, *J. Am. Chem. Soc.*, 102, 7196, 1980.

66. **Gavezzotti, A.**, The calculation of molecular volumes and the use of volume analysis in the investigation of structured media and of solid-state organic reactivity, *J. Am. Chem. Soc.*, 105, 5220, 1983.

67. **Connolly, M. L.**, Analytical molecular surface calculation, *J. Appl. Crystallogr.*, 16, 548, 1983.

68. **Marsili, M.**, Computation of volumes and surface areas of organic compounds, in *Proc. Beilstein Workshop Estimation of Physical Data for Organic Compounds*, Springer-Verlag, Berlin, 1988.

69. **Engler, E. M., Andose, J. D., and von Schleyer, P.**, Critical evaluation of molecular mechanics, *J. Am. Chem. Soc.*, 95, 8005, 1973.

70. **White, D. N.**, The principles and practice of molecular mechanics calculations, *Comput. Chem.*, 1, 225, 1977.

71. **Williams, J. E., Stang, P. J., and von Schleyer, P.**, Physical organic chemistry: quantitative conformational analysis; calculation methods, *Annu. Rev. Phys. Chem.*, 19, 531, 1968.

72. **Burkert, U. and Allinger, N. L.**, *Molecular Mechanics*, American Chemical Society, Washington, D.C., 1982.

73. **Sprague, J. T., Tai, J. C., Yuh, Y., and Allinger, N. L.**, The MMP2 calculation method, *J. Comput. Chem.*, 8(5), 581, 1987.

74. **Rasmussen, K.,** *Potential Energy Functions in Conformational Analysis,* Lecture Notes in Chemistry, Vol. 8, Springer-Verlag, Berlin, 1985.

75. **Murray, W., Ed.,** *Numerical Methods for Unconstrained Optimization,* Academic Press, London, 1972.

76. **Atkinson, K. E.,** *An Introduction to Numerical Analysis,* John Wiley & Sons, New York, 1978.

77. **Castellani, G. and Scordamaglia, R.,** A fast computer program for conformational analysis, *Comput. Chem.,* 8, 127, 1984.

78. **Scordamaglia, R. and Barino, L.,** New geometrical and electronic descriptors of molecule for structure-activity relationships, in *QSAR and Strategies in the Design of Bioactive Compounds,* Seidel, J. K., Ed., VCH Publishers, Weinheim, Federal Republic of Germany, 1985, 299.

79. **Fusco, R., Caccianotti, L., and Tosi, C.,** New algorithms to look for the most stable conformations of a molecule, *Nuovo Cimento,* 8, 211, 1986.

80. **Barino, L. and Scordamaglia, R.,** Conformation Analysis and Order-Disorder Transitions in Flexible Molecular Chains, Rolduc Polymer Meeting, Rolduc, The Netherlands, 1987.

81. **Clementi, E. and Sarma, R. H., Eds.,** *Structure and Dynamics,* Adenine Press, New York, 1983.

82. **Karplus, M.,** Molecular dynamics of biomolecules: overview and applications, *Isr. J. Chem.,* 27, 121, 1986.

83. **Levitt, M.,** Protein conformation, dynamics and folding by computer simulation, *Annu. Rev. Biophys. Bioeng.,* 11, 251, 1982.

84. **Heinzinger, K.,** MD simulations of the effect of pressure on the structural and dynamical properties of water and aqueous electrolyte solution, in *Supercomputer Simulations in Chemistry,* Dupuis, M., Ed., Springer-Verlag, Berlin, 1986, 261.

85. **Dupuis, M., Ed.,** *Supercomputer Simulations in Chemistry,* Springer-Verlag, Berlin, 1986.

86. **Del Re, G.,** Ground state charge transfer and electronegativity equalization, *J. Chem. Soc. Faraday Trans. 2,* 77, 2067, 1981.

87. **Mullay, J.,** A simple method for calculating atomic charge in molecules, *J. Am. Chem. Soc.,* 108, 1770, 1986.

88. **Abraham, R. J., Griffiths, L., and Loftus, P.,** Approaches to charge calculations in molecular mechanics, *J. Comput. Chem.,* 3, 407, 1982.

89. **Hinze, J. and Jaffe, H. H.,** Electronegativity. I. Orbital electronegativity of neutral atoms, *J. Am. Chem. Soc.,* 84, 540, 1962.

90. **Marsili, M.,** Ladungsverteilungen und Reactivität in der computergesteuerten Reactions-simulation, Ph.D. thesis, Technical University Munich, Munich, Federal Republic of Germany, 1980.

91. **Gasteiger, J. and Marsili, M.,** A new model for calculating atomic charges in molecules, *Tetrahedron Lett.,* 19, 3181, 1978.

92. **Gasteiger, J. and Marsili, M.,** Interative partial equalization of orbital electronegativity — a rapid access to atomic charges, *Tetrahedron,* 36, 3219, 1980.

93. **Basch, H., Viste, A., and Gray, H. B.,** Valence orbital potentials for atomic spectral data, *Theor. Chim. Acta,* 3, 458, 1965.

94. **Sanderson, R. T.,** *Chemical Periodicity,* Reinhold Publishing, New York, 1961.

95. **Sanderson, R. T.,** *Chemical Bonds and Bond Energy,* Academic Press, New York, 1976.

96. **Hinze, J., Withehead, M. A., and Jaffe, H. H.,** Electronegativity. Bond and orbital electronegativities, *J. Am. Chem. Soc.,* 85, 148, 1963.

97. **Huheey, J. E.,** The electronegativity of groups, *J. Phys. Chem.,* 69, 3284, 1965.

98. **Huheey, J. E.,** Group electronegativity and polar substituent constants, *J. Org. Chem.,* 31, 2365, 1966.

99. **Mortier, W. J., Ghosh, S. K., and Shankar, S.,** Electronegativity equalization method for the calculation of atomic charges in molecules, *J. Am. Chem. Soc.,* 108, 4315, 1986.

100. **Gasteiger, J. and Hutchings, M.,** Residual electronegativity — an empirical quantification of polar influences and its application to the proton affinity of amines, *Tetrahedron Lett.,* 25, 2541, 1983.

101. **Marsili, M. and Gasteiger, J.,** Pi-charge distributions from molecular topology and pi-orbital electronegativity, *Croat. Chem. Acta,* 53, 601, 1980.

102. **Gasteiger, J. and Saller, H.,** Berechnung der Ladungsverteilung in konjugierten Systemen durch eine Quantifizierung des Mesomeriekonzeptes, *Angew. Chem. Int. Ed. Engl.,* 24, 687, 1985; *Angew. Chem.,* 97, 699, 1985.

103. **Yanovskaya, L. A., Kryshtal, G. V., Yalovlev, I. P., Kucherov, V. F., Simkin, B. Y., Bren, V. A., Minkin, V. I., Osipov, O. A., and Tumakova, I. A.,** Transmission of electronic effects in certain polyenes, *Tetrahedron,* 29, 2053, 1973.

104. **Gasteiger, J. and Marsili, M.,** Prediction of proton magnetic resonance shifts: the dependence of hydrogen charges obtained by iterative partial equalization of orbital electronegativity, *Org. Magn. Reson.,* 15, 353, 1981.

105. **Gasteiger, J. and Guillen, M. D.,** Dipole moments obtained by iterative partial equalization of orbital electronegativity, *J. Chem. Res.,* S, 304, 1983; M, 2611, 1983.

106. **Guillen, M. D. and Gasteiger, J.,** Extension of the method of iterative partial equalization of orbital electronegativity to small ring systems, *Tetrahedron,* 39, 1331, 1983.

107. **Gasteiger, J., Marsili, M., and Paulus, B.,** Investigations into chemical reactivity and planning of chemical syntheses, in *Data Precessing in Chemistry,* Studies in Physical Theoretical Chemistry, Vol. 16, Hippe, Z., Ed., Elsevier, Amsterdam, 1981, 229.

108. **Koopmans, T.,** Ueber di Zuordnung von Wellenfunktionen und Eigenwerten zu den einzelnen Electronen eines Atoms, *Physica,* 1, 104, 1934.

109. **Gasteiger, J. and Hutchings, M.,** Empirical models of substituent polarisability and their application to stabilization effects in positively charged species, *Tetrahedron Lett.,* 24, 2537, 1983.

110. **Gasteiger, J. and Hutchings, M.,** Quantification of effective polarizability. Applications to studies of X-ray photoelectron spectroscopy and alkylamine protonation, *J. Chem. Soc. Perkin Trans. 2,* 3, 559, 1984.

111. **Gasteiger, J. and Hutchings, M.,** Quantitative models of gas-phase proton transfer reactions involving alcohols, ethers, and their thio analogs. Correlation analyses based on residual electronegativity and effective polarizability, *J. Am. Chem. Soc.,* 106, 6489, 1984.

112. **Mortier, W. J., Van Genechten, K., and Gasteiger, J.,** Electronegativity equalization: applications and parametrizations, *J. Am. Chem. Soc.,* 107, 829, 1985.

Chapter 5

AUTODEDUCTIVE SYSTEMS FOR REACTION KINETICS

I. INTRODUCTION

From this chapter on we present an in-depth treatment of some principal computerized systems that constitute the very essence of computer chemistry programs. Generally they are classified as autodeductive systems, being the result of a brilliant synthesis of AI-oriented programming and traditional chemical data processing.

Autodeductive systems are programs intended to assist the chemist in the design of experimental strategies, the main goal in computer chemistry. In such systems the computer fulfills the role of an expert adviser who can (1) think chemically, analyzing present system configurations; (2) predict future system configurations; and (3) reveal alternatives to established procedures in the laboratory.

Autodeductive systems work without preexisting mathematical assumptions; only the fundamental laws of conservation of mass and energy must be obeyed. The predictive facilities of such systems, in their most advanced realizations, do not use state-of-the-art libraries of prestored solutions to given problems: the solutions are deductively inferred by the computer after internal, user-independent analysis of the problem configuration.

Today, two types of autodeductive systems exist. The first type incorporates numeric autodeductive systems, and the second type employs alphanumeric or semantic autodeductive systems.

Numeric systems have both computative and predictive capabilities. The latter facility, typical of purely semantic systems, is used for the determination of combinations of parameters for which data are needed experimentally to compute numerical values for any other subset of system-pertinent parameters, or for predicting more complex entities (e.g., reaction educts from a specified target compound). In this last case the predicted entities are mostly of an alphanumeric nature, equivalent to chemical symbols.

Numeric systems also perform heavy mathematical evaluations, adding numbers to the logical predictive process. They can attribute weighted probabilities of existence as well as numerical values to the variables of a processed system — for example, the probability of obtaining specific reaction products from given educts (calculated from bond reactivity numbers) or the concentration of a certain reactant at a given time. Here the objects of the prediction are symbols and numbers. The distinction between the two kinds of systems currently is becoming less sharp, as the numeric performance of semantic programs is constantly increasing. The following section will focus on numeric autodeductive systems and will be exemplified by the presentation of one major achievement in this field.

II. PRINCIPLES OF NUMERIC AUTODEDUCTIVE SYSTEMS

In experimental chemistry, a research objective is generally pursued through the following steps:

1. Identification of the phenomenon to be studied and the goal
2. Hypothesis of a model and consequent tentative outline of procedures to establish a possible design for the experiment
3. Design or selection of the experiment(s) to be performed
4. Execution of the experiment(s) and collection of data
5. Data processing (by computers) or human data interpretation

6.   Draw conclusions (if data insufficient go to 3, if conclusions do not fit model go to 2)

Especially in industry, where this operational procedure is conditional on time, money, and, much worse, on each failure, strategies for global optimization of the experimental work are sought eagerly. An experiment can be optimized by acting on the minimum number of system variables sufficient to determine the whole system, avoiding any redundancy in the measurements. If several choices of subsets of variables for a given system are possible, then those which are easiest to measure and/or less costly to control should be selected.

One needs, then, an autodeductive system that helps in the selection of an optimal subset of system variables in order to allow the researcher to set up a modified, optimized working strategy. The strategic selection of subsets of variables, together with experimental data, should allow the computation of the values of the remaining variables under different system conditions. The system must have the following properties:

1.   It must have a user interface capable of accepting questions, possibly in an easily styled, semantic fashion (symbols), and also capable of returning answers in a chemical language.
2.   The system must be able to generate all combinations of input data (i.e., to conceive all possible strategies) that can be used to construct valid solutions in data processing step 5. This recalls the requirement of completeness for the program, which is therefore an autodeductive algorithm: all possible states of the problem space are detected. This predictive, strategic facility acts directly in decisional step 3.
3.   The system must have an automatic data processing capability for evaluating collected data in experiments designed with or without the predictive capability. There is a unique relationship between prediction and computation modules in a numeric auto-deductive system: the prediction module will provide all those combinations of the variable values for which it can derive a valid solution for the given problem, if input data for the computation module is provided.

III. THE CRAMS SYSTEM

The Chemical Reaction Analysis and Modeling System (CRAMS) was historically one of the first and also is one of the most powerful numeric autodeductive systems.[1,2] It is designed to help the chemist optimize the number of measurements in rate- or equilibrium-controlled chemical reactions. A chemical reaction is a system in which reagents are modified constantly, interacting with each other along specific reaction paths over a certain period of time. The system is described by two classes of parameters. The first class identifies parameters that vary during the chemical reaction, like concentration, pH, weight, density, color, volume, etc. The second class identifies constant parameters, like equilibrium or rate constants, that normally do not change during the experiment. If the reaction model belongs to the type for which a simple, unambiguous analytical solution exists, a direct application of such a known solution of the differential equations will be enough. For the reaction equation shown below, the solution is known.

$$A + B \rightarrow C; \quad k_c$$

In such a standard case, it is also quite simple to determine the minimum number of parameters necessary to calculate the values for the remaining system parameters. For example, the above reaction has the solution

$$k_c = (1/t) \ln[B^0(A^0 - x)/A^0(B^0 - x)]/(A^0 - B^0)$$

where A^0 and B^0 are the initial concentrations of A and B, respectively, and x is the concentration of C. Knowing k_c one can calculate x at any time. Conversely, knowing x at different times t, the reaction constant can be computed. This situation is conceptually very simple, as the question never arises about what could be measured and what could be left unmeasured, but computed. A totally different situation is given by the following network of reactions:

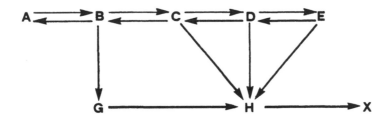

A perception of the solution structure is impossible, and it is not evident which sets of parameters can be computed from others. The following questions, then, are of interest to the researcher:

1. For which combinations (subsets) of known system parameters can the other parameters be calculated?
2. Once such subsets are known, how can the values for the residual parameters be calculated in practice?
3. Which parameters are always necessary to compute other parameters?
4. What is the validity of the computed parameters inside the postulated reaction model?
5. If the model seems unsatisfactory, can an alternative model be designed that better fits the measurements and the computed predictions?

Two main types of questions are found here: predictive questions (points 1 and 3) and computational question (points 2 and 4). Predictive questions seek information about the various combinations of parameters for which experimental data must be collected in order to compute the values of other parameters and to quantitatively test the goodness of the proposed reaction model. This is exactly the point where a great strategic impact is made on real laboratory work. The researcher can choose from all computer-proposed subsets of experimental parameters those that he regards as being the most convenient to manage. In addition, as we shall see in some examples, he becomes aware that only a fraction of the parameters are normally necessary to monitor the values of the residual ones. Thus, avoiding unnecessary measurements is a direct optimization of the overall experimental effort. Another way to formulate the question is: if it is very difficult (or impossible) to measure a particular parameter, what alternative parameter could one measure in order to attain the same final global information over the studied system?

Those solutions in which all parameters are either given or computed are called total solutions. Partial solutions contain at least one parameter that is neither given nor computed. Computational questions lead to the calculation of values for the constant and/or variable parameters and (if possible) to tests of the validity of the model. In processing computation questions, CRAMS first uses the predictive facility to determine the computable parameters and the correct order in which they must be derived. The computable parameters are obviously dependent on the user's input parameters, called the GIVEN parameters.

A. SEMANTIC INPUT

The input module in CRAMS is user friendly and highly semantic. The chemical equations are coded in the usual way and are augmented with indications to steer the program run into a predictive or computational mode. The core parts of the user interface are its SEMANTIC ANALYZER and PARSER, both of which are outfitted with error detecting and recovering procedures intended to guide the user to a correct use of CRAMS. PARSER checks the syntax and passes the parsed information on to the SEMANTIC ANALYZER. Incorrect names and improperly constructed "sentences" (e.g., an invalid form of a chemical equation) are detected during the semantic analysis.

The following reaction system illustrates a sample input:

$$A + B \underset{RK_2}{\overset{RK_1}{\longleftrightarrow}} C; \quad C \overset{RK_3}{\longrightarrow} A + D$$

CRAMS input deck:

/: EXAMPLE —— SIMPLE RATE REACTION SYSTEM :/

SYSTEM:
SELECT = 1
EQUATIONS:
A + B ↔ C, RK1, RK2; C → A + D, RK3;
CONSTANTS:
/: ALL CONCENTRATIONS ARE GIVEN OR COMPUTABLE :/
A = 0; B = 0; C = 0; D = 0;
/: ALL RATE CONSTANTS ARE GIVEN OR COMPUTABLE :/
RK1 = 0; RK2 = 0; RK3 = 0;
STOP;

SELECT = 1 calls for a predictive run. The initialization of all parameters with zero means that they are experimentally obtainable (GIVEN means user given) or computable. This decision comes from the user's knowledge: only he is in a position to decide whether a particular parameter (e.g., the reaction constant k_3) is easily measurable. If he has no particular constraints (of a technical or financial nature) concerning the procurement of the values for some parameters, these should all be flagged with GIVEN.

In a predictive run the GIVEN attribute does not involve a real number from a real experiment for every parameter in question, as no quantitative evaluation has been performed yet. Only the strategic level has been investigated so far, a level aimed precisely at defining which parameters should be measured later in the experiment!

B. PREDICTIVE QUESTIONS

From the semantic input an internal representation of the equation system is generated. Among the mathematical features present in CRAMS' predictive module, one deserves special attention: the FLUX matrix, which contains a concise description of the reaction model. Its elements are the molecularities of the reactants. Their positions in a matrix row are bound to the specific constants for the reaction in which the species are involved. For our example we have the equation system

$$d[A]/dt = -k_1[A][B] + k_2[C] + k_3[C] \tag{1a}$$

$$d[B]/dt = -k_1[A][B] + k_2[C] \tag{1b}$$

$$d[C]/dt = k_1[A][B] - k_2[C] - k_3[C] \tag{1c}$$

$$d[D]/dt = k_3[C] \tag{1d}$$

yielding the FLUX matrix

$$
\text{FLUX} = \begin{array}{c}
 \\
A \\
B \\
C \\
D
\end{array}
\begin{array}{|ccc}
\text{RK}_1 & \text{RK}_2 & \text{RK}_3 \\
\hline
-1 & 1 & 1 \\
-1 & 1 & 0 \\
1 & -1 & -1 \\
0 & 0 & 1
\end{array}
$$

The FLUX matrix is manipulated inside the SELECTOR part of CRAMS, which determines the parameters that can be computed and the single equations that have to be solved. This is achieved by manipulating the FLUX matrix with a special algorithm that involves iterative applications of Gaussian eliminations to determine the rank of comatrices in the FLUX matrix from which the values for unknown parameters can be computed. An important step is the determination of redundant information or, in other words, of linearly dependent equations. In fact, manipulation of Equations 1a through 1d shows that $d[A]/dt = -d[C]/dt$ and that $d[D]/dt = d[A]/dt - d[B]/dt$. SELECTOR integrates these equations, giving

$$[C] = C^0 - [A] + A^0$$

$$[D] = D^0 + [A] - A^0 - [B] + B^0$$

with A^0, B^0, C^0, and D^0 being the initial concentrations. Two differential equations have been simplified into two algebraic equations; two parameters, the concentrations of C and D, can be obtained from two others, A and B.

SELECTOR attends to other tasks of strategic relevance. In predictive questions, the GIVEN vector defining the status of each parameter (e.g., GIVEN, NONMEASURABLE, COMPUTABLE) is used to establish all possible complete and partial solutions. Every parameter flagged with GIVEN or COMPUTABLE is in turn regarded as user given, and the SELECTOR procedure traces all unknown parameters that can be calculated from the given parameters using the FLUX matrix procedure. This process is repeated with all possible combinations of GIVEN/COMPUTABLE parameters.

Similarly, NONMEASURABLE or NONCOMPUTABLE parameters are purged sequentially from the FLUX matrix in order to reduce it to a system containing only significant information. However, the computer issues warnings for NONCOMPUTABLE parameters in the final output of a CRAMS run; it must be understood by the reader that this information has a significant, informationally positive value for the experimenter, for he can readjust his measurements in order to collect data for other GIVEN parameters which permit the computation of the previously NONCOMPUTABLE ones. A revised strategy at an early stage is a form of experimental optimization.

C. COMPUTING QUESTIONS

In computing questions the results from SELECTOR are piped to the COMPUTATOR module, which computes values for the unknown computable variable parameters or values for the ordinates of the linear and nonlinear differential equations that must be solved for the computable unknown constant parameters. In there are three classifications of reaction systems that are accepted by SOLVER: equilibrium only systems, simulation only systems, and other types of systems.

Equilibrium only systems — SOLVER solves all nonlinear equations one at a time for the values belonging to all unknown computable parameters.

Simulation only systems — Such systems contain at least one rate equation, and only the concentrations of reactants have to be computed. The computation originates from a minimum set of independent equations, as determined in SELECTOR by a FLUX matrix manipulation. The calculational approach used can be summarized as follows: the minimal step length for the time range is computed, followed by a Kutta-Runge computation of concentration increments for the reactants and their derivatives. At each time step the system is forced to equilibrium by intervention of the SOLVER routine to compute the new equilibrium concentrations. The solution of differential equations calls for extended error-checking and error-recovery procedures. These procedures collectively ensure that only mathematically sound answers are formulated; in the cases where this appears impossible, the user is informed. It is up to him, then, to take counteractions before the job is resubmitted.

Other types of systems — In all other cases, the SOLVER program is responsible for generating values for the ordinates of the nonlinear (equilibrium only systems) and ordinary linear differential equations describing the reaction system constants. In these approaches, the derivatives are treated as ordinates and maximum tolerances for the dependent variables are evaluated as well.

The actual numerical computations are performed by the CURFIT system,[3,4] a CRAMS-integrated software package for automatic curve-fitting purposes. In CRAMS, the numerical integration of differential equations to solve for system constants is eluded. Instead, the derivatives are treated as other variable parameters. For example, the differential equation

$$d[A]/dt = -k_l[A][B]$$

is solved for k_l by fitting the experimental values [A] and [B] and the estimated derivatives $-d[A]/dt$ to a straight line. The so-called maximum tolerance (i.e., the uncertainty for every value of the dependent variable $-d[A]/dt$) must also be calculated to ensure stable fitting criteria. The maximum tolerances ultimately are used to recognize and reject "wrong" data points, to detect unsuspected curvatures (with the possible postulation of a different reaction model), and, eventually, to compute the maximum errors for the computed parameters. The derivatives are computed at each time for each variable in the following manner: three distinct couples of adjacent time-parameter values (t,x) are fitted to a second-order polynomial $(x = at + bt^2 + c)$. Differentiation gives the equation $dx/dt = a + 2bx$. The derivative for the interior values of x is computed initially, and the procedure is repeated until all derivatives are obtained except for the two terminal points. The terminal derivatives are calculated by fitting in turn the left and the right halves of the n available data points to a polynomial of degree $(n - 4)/2$.

IV. DESIGNING AN EXPERIMENT

A. EXAMPLE 1

The first example to demonstrate the performance of the CRAMS system deals with the strategic question of an optimal selection of experimental parameters for a given experiment, parameters from which others can be calculated (avoiding expensive additional measurements). For simplicity we consider the reaction system described above, for which a CRAMS input has already been formulated. Here the user is solely interested in a strategic problem, and the parameters are treated as symbols: no real numbers are given in the input deck. PREDICTOR analyzes the problem, revealing which parameters can be computed (C in the output) and which of them cannot be computed (blank in the output) from a specific initial subset of given parameters (G in the output). Table 1 lists all possible complete and partial solutions.

TABLE 1

The Predictive Answers Given by CRAMS for the Equation System A + B → C,
C→ A + D

14 NONREDUNDANT PREDICTIONS — ENTRIES ARE AS FOLLOWS
(G — GIVEN; R — RECOMPUTED; BLANK — NOT COMPUTED;
N — NOT MEASURABLE; C — COMPUTED; NC — NOT MEASURABLE AND COMPUTED)

Name	1	2	3	4	5	6	7	8	9	10	11	12	13	14
A	G	G	G	G	G	C	C	C	C	C	C	C	C	C
B		G	C	C	C		G	G	G	C	C	C	C	C
C	C	C	C	C	C	G	G	C	C	G	G	G	C	C
D		C	G	C	C		C	G	C	G	C	C	G	C
RK1		C	C	G	C		C	C	C	C	G	C	G	G
RK2		C	C	G	C		C	C	C	C	G	C	G	G
RK3		C	C	C	G		C	C	G	C	C	G	C	G
	1	2	3	4	5	6	7	8	9	10	11	12	13	14

Partial solutions 1 and 6 are trivial, while the remaining twelve solutions demonstrate two major points. First, complete solutions exist only if at least one of the following pairs of system parameters are actually measured: A,B or A,D or B,C or C,D. A strategic consequence would be to consider the time and the financial and technical effort necessary to collect data for the different pairs of variables and plan your experiment accordingly.

Pair A,C does not give a solution (there is no simultaneous presence of G symbols for A and C in one and the same column). A strategic consequence would be to avoid setting up an experiment monitoring the concentrations of A and C.

Second, some complete solutions require the determination of three parameters. A strategic consequence would be to plan your experiment to involve the lowest possible number of measurements. However, additional data (and the computer tells us which data in every case) can help in attesting a proposed reaction model.

B. A COMPUTATIONAL EXAMPLE

To illustrate a computing problem, we can alter the previous input into the following:

```
SYSTEM:
SELECT = −1
EQUATIONS:
A + B ↔ C, RK1, RK2; C → A + D, RK3;
CONSTANTS:
/: ALL RATE CONSTANTS ARE KNOWN :/
RK1 = 0.001; RK2 = 10.0; RK3 = 0.10;
/: INITIAL CONCENTRATIONS :/
A = 0.004; B = 1.0;
DATA:
TIME:
     1.0 1.5 2.0 2.15 5.0;
STOP;
```

The problem we are faced with here is, given the indicated values for the reaction constants and the initial concentrations of the reactants, to calculate concentrations for all reactants at the given times. The initial (default) concentration for both C and D is zero. SELECT = −1 indicates a computational run. The results shown in Table 2 are obtained.

TABLE 2
Concentrations at Given Times Computed by CRAMS for the Four Reaction Components A, B, C, and D

4 SIGNIFICANT FIGURES ARE TO BE PRINTED FOR EACH OF THE
5 VALUES FOR EACH OF THE 4 COMPOUNDS.

Time	A	B	C	D
0.1000D 01	0.4000D-02	0.1000D 01	0.0	0.0
0.1500D 01	0.4000D-02	0.1000D 01	0.3935D-06	0.1590D-07
0.2000D 01	0.4000D-02	0.1000D 01	0.3960D-06	0.3558D-07
0.2150D 01	0.4000D-02	0.1000D 01	0.3960D-06	0.4162D-07
0.5000D 01	0.4000D-02	0.1000D 01	0.3960D-06	0.1545D-06

TABLE 3
The Complete Set of Predicted Solutions for the Equilibrium System
A ↔ B, B ↔ Z

10 NONREDUNDANT PREDICTIONS — ENTRIES ARE AS FOLLOWS
(G — GIVEN; R — COMPUTED; BLANK — NOT COMPUTED;
N — NOT MEASURABLE; C — COMPUTED; NC — NOT MEASURABLE AND COMPUTED)

Name	1	2	3	4	4	6	7	8	9	10
A	G	G	G	G	C	C	C	C	C	C
B	G	C	C	C	G	G	G	C	C	C
Z	C	G	C	C	G	C	C	G	G	C
EK1	C	C	G	C	C	G	C	G	C	G
EK2	C	C	C	G	C	C	G	C	G	G
	1	2	3	4	5	6	7	8	9	10

C. AN EQUILIBRIUM SYSTEM

With PREDICTOR the following equilibrium reaction system must be studied:

$$A \overset{EK_1}{\longleftrightarrow} B; \qquad B \overset{EK_2}{\longleftrightarrow} Z$$

where EK_1 and EK_2 are the equilibrium constants. The input is formulated in the following way:

/: EXAMPLE 2 : SIMPLE EQUILIBRIUM REACTIONS :/
SYSTEM;
SELECT = 1;
EQUATIONS:
/: FIRST EQUILIBRIUM REACTION :/
A ↔ B, EK1;
/: SECOND EQUILIBRIUM REACTION :/
B ↔ Z, EK2;
CONSTANTS:
/: SET ALL PARAMETERS TO GIVEN OR COMPUTABLE :/
A = 0; B = 0; Z = 0; EK1 = 0; EK2 = 0;
STOP;

The simulation yields ten nonredundant complete solutions, as shown in Table 3. All

TABLE 4
Computational Results for the Equilibrium System Obtained by Solution #8 of the Prediction Run

Given a known preequilibrium concentration of educt A and the value for EK1, CRAMS computes the concentration of the other added components (A and B) and the time required to reach equilibrium

THE COMPUTATIONAL STATUS OF ALL PARAMETERS IS GIVEN NEXT.
4 SIGNIFICANT FIGURES WERE REQUESTED FOR THE CONSTANT PARAMETERS.

A	WAS COMPUTED.
B	WAS COMPUTED.
Z	WAS GIVEN.
EK1 = 0.5000D 00	WAS GIVEN.
EK2 = 0.1000D 01	MAXIMUM ERROR = 0.0 WAS COMPUTED.

4 SIGNIFICANT FIGURES ARE TO BE PRINTED FOR EACH OF THE 4
VALUES FOR EACH OF THE 3 COMPOUNDS.

Time	A	Z	B
0.1000D 01	0.3000D 01	0.1500D 01	0.1500D 01
0.2000D 01	0.5000D 01	0.2500D 01	0.2500D 01
0.3000D 01	0.7000D 01	0.3500D 01	0.3500D 01
0.4000D 01	0.9000D 01	0.4500D 01	0.4500D 01

parameters can be computed from a suitable subset of measured parameters. Prediction 8 informs us that all other unknown parameters can be calculated if the concentration of Z and the value of EK_1 are known. Inserting solution 8 into CRAMS using the input specifications listed below, one obtains the equilibrium concentrations for A and B and the value for EK_2.

```
SYSTEM;
EQUATIONS:
A ↔ B, EK1; B ↔ Z, EK2;
CONSTANTS:
EK1 = 0.5;
/: GIVE INITIAL CONCENTRATION :/
INITIALC:
A = 1;
/: GIVE INITIAL PREEQUILIBRIUM CONCENTRATIONS :/
REINITIAL:
B, Z;
2 3 4 5 6 7 8 9; /: MAKE FOUR RUNS :/
DATA:
TIME, Z;
/: Z HAS FOLLOWING EQUILIBRIUM CONCENTRATIONS :/
0 1.5 0 2.5 0 3.5 0 4.5;
STOP;
```

The computational results are displayed in Table 4. The TIME values inform the user about the time required to reach an equilibrium state in each of the four investigated situations.

D. A MORE COMPLEX EXAMPLE

The examples just presented have an evident demonstrative character, tending to highlight the principles for the use of CRAMS. This system also can handle very intricate reaction networks for which the human mind has no chance to identify possible optimized solution schemes. The following input is an example of a predictive run processing a fairly complicated reaction system. The rate constants have "F" and "B" extensions for forward and backward reactions, respectively.

```
SYSTEM:
SELECT = 1;
EQUATIONS;
/: RATE EQUATIONS FIRST :/
EHA ↔ EI, K1F, K1B;
EI ↔ EC, K2F, K2B;
EC ↔ EHA, K3F, K3B;
EH ↔ EP + H, KHF, KHB;
E ↔ EP, KEF, KEB;
/: EQUILIBRIUM EQUATIONS :/
EI ↔ E + I, KI;
EHA ↔ EH + A, KHA;
EC ↔ E + Z, KC;
CONSTANTS:
/: CONSTANTS MARKED WITH 1 ARE GIVEN :/
K1F = 1; K1B = 1;
K2F = 1; K2B = 1;
K3F = 1; K3B = 1;
KI = 1;
KHA = 1; KHF = 0; KHB = 0;
KC = 1;
EHA = 0; EI = 0; EC = 0; EH = 0; H = 0; E = 0; I = 0;
A = 0; Z = 0;
STOP;
```

The self-explanatory predictive results are listed in Table 5. Note that the chemist investigating this complicated reaction network cannot measure the constants KEF and KEB for whatever reasons. Nonetheless, they can be computed by the CRAMS system.

TABLE 5
A Printout of the Complete Predictive Set of Solutions for the Complex Reaction System Given Above

19 NONREDUNDANT PREDICTIONS – ENTRIES ARE AS FOLLOWS
(G – GIVEN; R – RECOMPUTED; BLANK – NOT COMPUTED; N – NOT MEASURABLE;
C – COMPUTED; NC – NOT MEASURABLE AND COMPUTED)

Name	1	2	3	4	5	6	7	8	9	10	11	12	13	14	15	16	17	18	19
EHA	G	G	G	G	C	C	C	C	C	C	C	C	C	C	C	C	C	C	C
EI	G	C	C	C	G	G	G	G	C	C	C	C	C	C	C	C	C	C	C
EC	C	G	C	C	C	C	C	C	G	G	G	G	C	C	C	C	C	C	C
EH	C	C	C	C	G	C	C	C	C	C	C	C	G	G	C	C	C	C	C
EP	C	C	G	C	C	G	C	C	C	G	C	C	G	C	G	G	G	C	C
H	C	C	C	C	C	C	G	C	C	C	G	C	C	C	G	C	C	G	C
E	C	C	C	G	C	C	C	C	C	C	C	C	C	G	C	G	C	G	G
I	C	C	C	C	C	C	C	C	C	C	C	C	C	C	C	C	C	C	C
A	C	C	C	C	C	C	C	C	C	C	C	C	C	C	C	C	C	C	C
Z	C	C	C	C	C	C	C	C	C	C	C	C	C	C	C	C	C	C	C
KIF	G	G	G	G	G	G	G	G	G	G	G	G	G	G	G	G	G	G	G
KIB	G	G	G	G	G	G	G	G	G	G	G	G	G	G	G	G	G	G	G
K2F	G	G	G	G	G	G	G	G	G	G	G	G	G	G	G	G	G	G	G
K2B	G	G	G	G	G	G	G	G	G	G	G	G	G	G	G	G	G	G	G
K3F	G	G	G	G	G	G	G	G	G	G	G	G	G	G	G	G	G	G	G
K3B	G	G	G	G	G	G	G	G	G	G	G	G	G	G	G	G	G	G	G
KHF	C	C	C	C	C	C	C	G	C	C	C	G	C	C	C	C	G	C	G
KHB	C	C	C	C	C	C	C	G	C	C	C	G	C	C	C	C	G	C	G
KEF	NC	NC	NC	NC	NC	NC	NC	NC	NC	NC	NC	NC	NC	NC	NC	NC	NC	NC	NC
KEB	NC	NC	NC	NC	NC	NC	NC	NC	NC	NC	NC	NC	NC	NC	NC	NC	NC	NC	NC
KI	G	G	G	G	G	G	G	G	G	G	G	G	G	G	G	G	G	G	G
KHA	G	G	G	G	G	G	G	G	G	G	G	G	G	G	G	G	G	G	G
KC	G	G	G	G	G	G	G	G	G	G	G	G	G	G	G	G	G	G	G
	1	2	3	4	5	6	7	8	9	10	11	12	13	14	15	16	17	18	19

REFERENCES

1. **DeMaine, P. A. D.,** Automatic deductive systems for chemistry, *Anal. Chim. Acta,* 133, 685, 1981.
2. **DeMaine, P. A. D. and DeMaine, M. M.,** Automatic deductive systems. I. Chemical reaction models, *Comput. Chem.,* 1, 49, 1987.
3. **DeMaine, P. A. D. and Springer, G. K.,** A non-statistical program for automatic curve-fitting to linear and non-linear equations, *Manage. Inf.,* 3, 233, 1974.
4. **DeMaine, P. A. D.,** Operation and Logic Manual for the CURFIT System, Automatic Systems for the Physical Sciences Ser. Rep. 2, Computer Science Department, Pennsylvania State University, University Park, 1976.

Chapter 6

STRUCTURE ELUCIDATION SYSTEMS

I. GENERAL PRINCIPLES OF COMPUTER-ASSISTED STRUCTURE ELUCIDATION

In the previous chapter we discussed a preliminary example of an autodeductive system for designing chemical experiments. The essential contribution of self-deciding computer chemistry programs, we shall stress again, lies in their capability of presenting to the investigator a set of complementary solutions for a specific problem. These solutions belong to the problem space spanned by the system variables and must be mathematically and logically stable. Within the problem space boundaries the maximum possible number of solutions must be generated. This is a mandatory feature of any AI-oriented computer chemistry program: it must generate, in principle, all possible goal states from an initial state. "In principle" refers to the intrinsic, algorithmic capacity of the program. If, for example, the nature of the problem is of untractable combinatoric vastness, the complete set of solutions may never be created due to CPU time limits, to core memory restrictions, or, more likely, to an impendent unmanageable avalanche of solutions. Theoretical and heuristic restrictions must therefore intervene to select classes of solutions. Fortunately, in real situations not all of the goal states are requested by the investigator, who is mainly interested in either one specific goal state (i.e., in problems where only one solution exists *per definitionem*) or in the collection of a particular set of solutions. Conceptually, a distinction between real goals and virtual goals must be recognized: virtual goals are those that can be generated mechanistically by the computer, assuring completeness for the set of solutions; real goals are those among all virtual goals which, after passing a discriminatory testing phase, are judged to be consistent with the experimental findings.

A common chemical problem is the determination of the structure of an unknown molecule. The usual procedure involves the interpretation of a wealth of spectral data; frequently, the search for similar spectra in computerized spectral data bases is a welcome help for the laboratory chemist.

Structure elucidation is a combinatoric/semantic problem. A combinatoric problem dealing with a finite number of atoms has a finite number of solutions. Interpretation of chemical and spectral data forces our brains to make rational conclusions and sudden intuitive breakthroughs (which are not possible in a computer). This tactical approach, so familiar to all organic chemists, clearly involves the acquired knowledge of the chemist, his own personal cerebral data base and his feeling for analogy, through which he proceeds in the evaluation of the experimental data. If the elucidation problem is nontrivial from a combinatoric point of view, the researcher often finds the correct solution only after time-consuming labor (excluding trivial cases, obviously). The well-known fact that a chemist directs his attention, even at an early stage of his investigation, toward a class of structures that presumably includes the unknown structure is due to the selective way in which his "chemical" brain works. From the beginning he discards all those improbable structures which are in contrast to his experimental data after having conceived them in his mind. To better understand this point, let us consider the following simple substructures, $-CH_3$, $-OCH_2-$, $-OH$, and $-C_6H_4$, and attempt to assemble them. Using minimal formal reasoning, there are several topologically valid ways of mutually linking the substructures. For the chemist it will be quite easy to deduce the right structure, which should be uniquely identifiable from NMR and reactivity data. He will never be tempted to conceive a structure like

$$CH_3-C_6H_4-CH_2-OOH$$

if the reactivity of the unknown molecule points to an inert species. Intuitively, supported by his partial knowledge and by thinking in analogies, he conceives structures that are consistent with his expectations about the chemical nature of the unknown.

The trivial assembly example just discussed leads to other quite different classes of organic compounds, even if the four substructures are of a very simple nature. There are problems (e.g., in the structure elucidation of a natural product or of an unsuspected reaction by-product) that are difficult to solve. Frequently, several solutions which are all apparently in accordance with the spectral data are conceivable. The more underdeterminate the problem (i.e., the fewer substructures that can be matched unambiguously to spectral data), the larger the number of possible constitutive structures. The smaller the number of atoms included in recognized and validated substructures, the larger the number of unassigned free atoms needed to complete the molecular formula. For instance, if the molecular formula of an unknown structure was $C_{12}H_{22}N$ and all atoms of the identified substructures together gave a partial sum like C_6H_{10}, then 6 carbons, 12 hydrogens, and 1 nitrogen atom would still have to be incorporated into the final structure. Unfortunately, such unassigned atoms with their free valences give rise to a large number of structural isomers. The fantasy and intuition of the human chemist may be lost if he is confronted with a combinatorial explosion. He may no longer be able to construct the correct final structure within an acceptable period of time. In Chapter 1 a sort of chemical game existed in the attempt to conceive all isomeric structures of benzene. It might be surprising to the reader, but there are 217 of them (all different), quite a large number considering that benzene has only six carbon atoms! But if we are given the information that two methyl groups must be present in the final structure, only seven solutions remain:

The availability of computers has greatly facilitated the process of structure elucidation due to three main factors:

1. The dramatic time reduction in the solving phase
2. The guarantee that the complete set of possible structures is generated by the machine — the right one must be among them
3. The widespread spectrometric equipment, ensuring a large amount of computer-interpretable data for reliable automatic structure elucidation

A program that has been designed to perform this task is called a Structure Elucidation System (SES). Such programs are a milestone in the history of computer chemistry, and their realization has been strongly influenced by AI concepts and methodologies.[1]

An SES does not rely on a knowledge base of finished solutions, but it does depend uniquely on autodeductive properties. The solutions are constructed from chemical rules following the generate and test paradigm, which was briefly discussed in Chapter 3. These programs consequently belong to the class of semantic predictive systems. Their output consists of the direct display of complete molecular structures.

The strategic role of an SES is twofold: first, it does not suffer from "loss of memory", meaning that it will not forget structures, as can easily happen to a human. Second, it stimulates the chemist to reconsider his own structure proposal while confronting him with computer-generated structures: the visual perception of suggested structures is often a springboard to a novel enlightenment leading to a different interpretation of the spectral and chemical data. Both actions result in a direct optimization of laboratory work. Investments in time, in human effort, and in money are thereby reduced.

A. THE PLANNING PHASE

The following scheme illustrates the fundamental strategy characterizing an SES:

PLANNING $\longrightarrow$ **GENERATING** $\longrightarrow$ **TESTING**

This basic SES architecture is philosophically similar in all functioning SES programs. The implementations to be discussed later differ only in their actual specific tactical realization or in the kind of data processed. They are not divergent from the given strategic line and are all appreciated for their brilliant and pioneering synthesis in AI-oriented programming and chemical science.

The task of transforming a multiform amount of chemical and spectral data into a final correct structure is difficult, even for a computer, for underdeterminate problems of high combinatoric complexity. The problem can be simplified by the user's intervention, interactively introducing more spectral data and/or more chemical information. This mainly happens during the planning phase.

During this first operating stage the principal action consists of defining positive constraints. Positive constraints are substructures which are known to be present in the final structure one or more times. The perception of such constraining substructures can be realized in two ways:

1. The SES is informed by the chemist about such substructures; they are physically input in their explicit form during the interactive elucidation session.
2. The SES interprets raw spectral data (no previous human interpretation is necessary) and selects a set of compatible substructures; this is a totally automatic autodeductive SES.

A combination of methods 1 and 2 is very common in practice.

A fully automatic generation of constraints is achieved through a so-called spectrum interpreter. For example, interpreters can analyze the raw data of MS, IR, ^{1}H-NMR, ^{13}C-NMR, and 2-D NMR spectra and, by special recognition routines, derive a set of particular fragments that are in accordance with the spectral signals. The sum of the atoms of the ensemble of accepted substructures may or may not be equal to the molecular formula. If not, a certain number of still unassigned atoms remain. Every human-inferred piece of information is extremely valuable at this point. Any user-known substructure can reduce the amazing number of structures to be generated later enormously if introduced into the SES.

Chemical data are normally evaluated by the user; if, for example, halogen addition to double bonds occurs in the unknown molecule, then substructure C=C can be reasonably assumed to be present and is introduced into the computer.

There are also working prototypes of computerized chemical interpreters; they are utilized both in the planning and the final testing phases. They can accept (1) direct chemical information or (2) a coded chemical reaction. Information of the first type serves to create additional substructural information at the beginning of the SES run. In one system to be presented below (CASE), a special routine accepts the number of moles of periodate consumed in an oxidation reaction of the unknown. Information of the second type consists of a user-coded instruction addressed to the SES. It performs a specific simulated chemical reaction on the generated structures coming from the GENERATE phase. The virtual products are analyzed substructurally and matched with experimentally determined substructural features of the product(s) of the real reaction involving the unknown molecule. As preexisting structures are required to exploit the performance of such chemical interpreters, they find application in the testing stage, where the computer-generated structural candidates are scrutinized according to appropriate selection rules.

We can expand the SES strategy scheme as shown in the following diagram (POS. CON. = positive constraints):

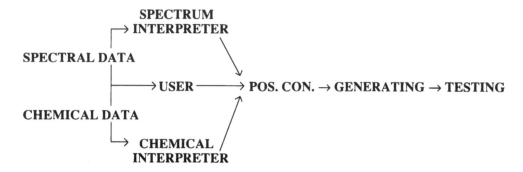

All data gathered during the PLANNING phase (i.e., the set of acceptable substructures and the molecular formula [user given or computer predicted; there are programs that compute the molecular formula from high-resolution MS][2]) are conveyed to the second phase — the GENERATING step.

B. THE GENERATING PHASE

In the GENERATING part of an SES the collected substructural information is processed by a molecule assembler, also called a structure generator. According to the free valences of the available substructures and of the free atoms (if any exist), all possible topological isomers are generated.

The structure generator contains a routine that guarantees the generation of all possible isomers (the solution candidates) which are in accordance with the constraints. Molecules that do not contain the required pattern of given substructures are never created. This mechanism effectively counters the threat of a combinatorial explosion.

Another routine avoids the storage of duplicates, thereby pruning redundant candidates. Structural redundance is recognized at once through a canonical numbering algorithm that uniquely names the generated structures. Equal names mean equal structures, which can then be deleted.

After generation of all possible candidates, the user can inspect them on a printout or have them displayed on a graphics screen. In fortunate cases where the initial spectral and chemical information was exhaustive, the number of candidates is reduced to just one final structure: the target structure of the unknown. Otherwise, and more frequently, a set of candidates survives which is not further discriminable with the available data. Human intervention can be strategic in the further reduction of candidates. By visual inspection the chemist can often detect structures containing particular substructural features that cannot be present in the real molecule. The negation of specific substructures, the negative constraints, is the result of human knowledge and judgment. Experience with the reactive behavior of investigated species, careful evaluation of the available spectral data, and confrontation with preceding analog compounds may persuade the investigator about the absence of some computer-generated substructures in the unknown.

Improbable or impossible substructures are not the result of an algorithmic error in the SES, but originate from the degrees of freedom inherent in the free, unallocated atoms and/ or in some given small substructures during the action of the molecule assembler. The spectral frequency range of a newly generated substructure, born in the GENERATING module from free atoms and now embedded in one or more candidates, may be incidentally similar to the spectral range of a positive constraint which is part of the validated input in the PLANNING step. As the new substructure is apparently not in spectral contrast to the input data, it becomes temporarily accepted, even in an SES where an early check by spectrum-simulating modules is used to monitor the generating process. More frequently, in the absence of such a checking module, two (or more) very small substructures activated during the PLANNING phase can give rise to a number of other, larger substructures of a highly contrasting chemical nature.

For example, if the structure generator has to deal with two $-CH_2N<$ substructures during the assembling phase, it can generate the alternatives

Substructures A and C are both amines, but B belongs to the class of diazo compounds and D to the diaziridine class. It will be evident from elementary chemical and chromatic characteristics of the unknown species whether or not these substructures are present. If no evidence for such groups is found by the chemist, he can turn this valuable information over to the SES as a negative constraint.

Inside the PLANNING stage a determination of positive constraints is carried out, whereas in the GENERATING phase the inclusion of negative constraints is of high strategic value for cutting down the number of candidates.

This mechanism of manual input of user-recognized negative constraints was the predominant working method in the early stages of SES development. An SES was mainly seen as a structure generator, and a great deal of human intervention was necessary to converge on a reasonable number of final candidates. This was the guiding philosophy of the SES CONGEN,[3,4] one of the first excellent software systems introducing automated chemical reasoning. CONGEN is one large program inside an entire important system of general structure elucidation programs developed within the DENDRAL project.[4]

Later approaches to the evaluation of structural candidates have been designed in view of an extensive processing of spectral data, greatly reducing human intervention.

The GENERATING phase thus can be propounded in the following diagram:

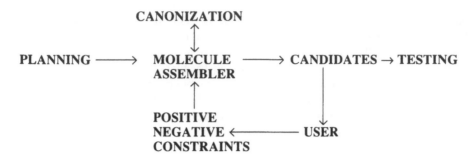

C. THE TESTING PHASE

The final step of an SES data processing routine is the TESTING phase. During this last phase, each surviving candidate undergoes a test determining its probability of being the correct structure. The test frequently consists of the simulation of one particular kind of spectrum (e.g., MNR, MS) of the actual candidate. The simulated spectrum is produced by a spectrum simulator inside the TESTING module. The predicted spectrum is then compared with the experimental spectrum of the unknown compound (part of the input in PLANNING), and a similarity index is computed between the two spectra; this action is called the matching procedure. Each candidate receives its own similarity index; the candidates are ranked according to their individual index magnitudes. Candidates with high indices are retained; those with low indices are deleted from the array of candidates. If symmetry properties of the candidates are used for simulating the number of peaks of their ^{13}C-NMR spectra, the diminution of candidates becomes dramatic: all candidates showing a number of predicted peaks different from the experimentally determined number are deleted. Another testing method has been mentioned already and consists of performing simulated reactions on the surviving candidates. This approach is exemplified in detail later in this chapter.

We can conclude the description of the general framework of an SES by expanding the TESTING step into its methodological functions:

$$\text{PLANNING} \rightarrow \text{GENERATING} \rightarrow \text{SPECTRUM SIMULATOR} \rightarrow \text{MATCHING} \quad \substack{\text{EXPERIMENTAL} \\ \text{SPECTRUM} \\ \updownarrow}$$

EXPERIMENTAL
SPECTRUM
⇕

PLANNING → GENERATING → SPECTRUM   → MATCHING
                         SIMULATOR   ↓
                                      RANKING
                                        ↓
                                      FINAL
                                      STRUCTURES

II. STRUCTURE GENERATION

Structure-generating algorithms are undoubtedly among those which have characterized the birth of computer chemistry and of structure elucidation systems. The early processes of structure assembly have been based on a stepwise expansion of the molecular graph, through which a substructure is gradually enlarged to form a complete final structure.

The assemblage starts each time from a universal set of all conceivable bonds that could be made to form a complete molecular structure. As a valid candidate must contain all positive constraints included in the problem space, a check for their presence in the generated structure must follow its completion. This, however, cannot be ascertained until the last bond of the complete structure has been formed. Candidates recognized with missing substructures will be excluded retrospectively.

This methodology strictly follows the generate-and-test strategy, according to which all problem states (candidate structures) are generated first and then analyzed for possible incongruity with the input constraints.

It follows that an improvement of the generating process can be gained in the structure reduction methodology,[5] as was demonstrated recently. It starts from the set of all possible bonds, but in contrast to the assembly method each of these possible bonds is initially made involving all substructures available to form a hyperstructure. Structure reduction proceeds with the progressive deletion of bonds inside the hyperstructure until a complete structure is obtained. The absence of a required substructure now is monitored prospectively. As soon as a first bond of a given substructure is deleted in the hyperstructure, the substructure itself becomes deformed. Structure generation along this path can be stopped immediately, avoiding futile generation of illicit candidates.

A classic algorithm for constrained structure generation, CONGEN, is worthy of receiving a closer look. It will be discussed here from its prominent aspect of being a generator, meaning that we shall concentrate our attention on the algorithmic part dealing with the generation of the constitutive molecular graphs, disregarding any consideration concerning constraint checking.

A. DEFINITIONS
The following technical and semantic definitions are necessary:

1.   Concepts of *saturated* and *unsaturated* compounds: a molecule is defined as saturated when it contains the maximum possible number of hydrogen atoms. A molecule is classified as unsaturated when this number is smaller than the maximum possible number. It follows from this agreement that cyclic compounds (cyclohexane, decaline, etc.) belong to the class of unsaturated compounds.

2.   The concept of *cycle*: every unsaturated molecule contains at least one cycle. Cycles are therefore not only the obvious rings (represented by some geometric symbol, like a triangle, square, pentagon, hexagon, etc.), but also include all species having at least one degree of unsaturation (ethylene, acetylene, etc.).

3.   A *symbolism* for unsaturation: a coded molecular formula with explicit inclusion of the degree of unsaturation is introduced. A capital U indicates a degree of unsaturation. Cyclohexane, C_6H_{10}, thus can be coded into C_6U_1, where U_1 already implicitly contains the necessary hydrogen atoms to complete the molecular formula conveniently. The reader should identify the structures belonging to C_2U_2, C_3U_2, and C_6U_4.

4.   The concept of the *superatom*: this is one of the prime concepts in CONGEN. A superatom represents any connected graph, which is treated as a single "atom" during the structure assembly. There are superatoms representing cycles and others marking acyclic substructures. For example, a phenyl group may be defined as superatom **A** and an acetyl group as superatom **B**. When assembling **A** and **B** one obtains the metastructure **A–B**. The expansions of the superatoms into the real structure (imbedding) leads to the formula $C_6H_5COCH_3$.

A superatom can have one or more free valences that are processed individually in the assembly phase. Let **C** represent a trisubstituted benzene unit, C_6H_3; **C** is then a trivalent superatom. If three other monovalent superatoms, **D**, **A**, and **B**, are joined to **C**, a tree of superatoms is obtained

which (if **C** is imbedded) yields the hybrid structures of Figure 1.

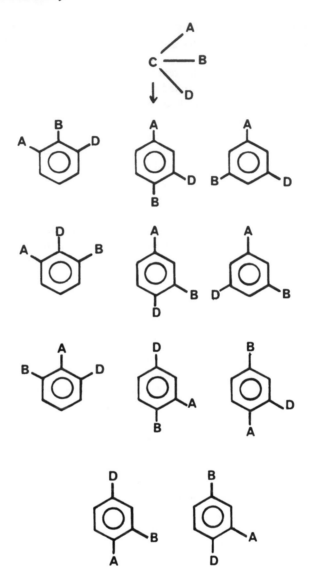

FIGURE 1.   Imbedding of a trivalent superatom **C** (trivalent benzene unit) linked to three monovalent superatoms (**D, E, F**).

The advantage of using superatoms in the generating step is given by the con-
spicuous simplification of the molecular graph: each superatom containing a cycle
condenses a substructural graph (a tree in acyclic substructures) into a single node of
the molecular graph.

If all cycles of a molecular structure could be condensed into an equivalent series
of superatoms, the molecular graph would be transformed into a molecular tree. The
canonization now acts on a cycle-free topological structure. The generation of a canonical, complete set of candidates is equivalent to the creation of all isomeric trees,
some nodes of which are superatoms.

5.   The concepts of *vertex graph* and *vertex atom*: A vertex graph is an abstract structure
pertaining to a specific class of superatoms. To every cycle-encoding superatom a
specific kind of vertex graph is related which contains the same number of cycles as
the superatom. A vertex graph defines an equivalence class of cycle-encoding superatoms.

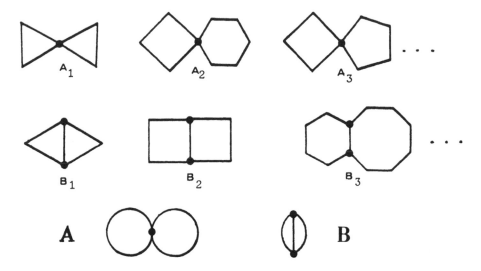

FIGURE 2. Classification of vertex graphs.

A vertex atom is an atom member of two or more cycles. Cycles showing only one vertex atom are called spirofused; cycles having at least two adjacent vertex atoms are called fused.

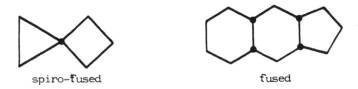

spiro-fused   fused

All superatoms with one tetravalent vertex atom (four edges departing from the node) belong to the class of one-tetravalent-vertex graphs, those with two trivalent vertex atoms to the class of two-trivalent-vertex graphs, and so on. Note the importance of the number of edges irradiating from the vertex atom: it corresponds to the number of bonds engaged in the real molecule. The total valence of the vertex atom minus the number of edges gives the number of residual free valences. For carbon atoms, vertex atoms with four edges do not possess free valences. The classification of superatoms by means of their vertex graphs conforms to the number of vertex atoms and their number of edges. How many "real" atoms are contained in the cycles of the respective superatoms is of no importance. For example, as shown in Figure 2, the superatoms encoding the structure A_1, A_2, A_3, ... and B_1, B_2, B_3,... belong to the two classes schematically represented by vertex graphs A and B, respectively. A catalog of thousands of chemically valuable vertex graphs is stored in the DENDRAL SES.

With these definitions we can proceed to the discussion of the generating algorithm.

B. THE GENERATING ALGORITHM

The CONGEN algorithm for the creation of all topological isomers consistent with a given molecular formula and a certain number of constraints follows an assemblage philosophy, and it is hierarchically composed as follows:

1.   From the globality of nonmonovalent atoms of the molecular formula, generate all

possible partitions in two sets. The first set, SSET, contains atoms forming cycle-encoding superatoms; the second set, RSET, contains all other atoms.

2.   For every generated partition, determine the class of vertex graph to which the SSET under consideration belongs.

3.   For every vertex graph class, generate all possible cycle-encoding superatoms. If a superatom violates a constraint, delete it.

4.   For every combination of cycle-encoding superatom and of atoms of the RSET (for a specific partition), generate with the acyclic generator all possible nonredundant molecular trees. Repeat this for each partition.

5.   For imbedding, expand every superatom node into the related full chemical substructure.

6.   Delete final structures in contrast with the given constraints.

The molecular formula is first translated into a notation evidencing the degree of unsaturation. For example, for $C_9H_{15}OF$, with two degrees of unsaturation, one can write C_9OU_2. Next, the possible partitions into SSET and RSET are envisaged. It must be remembered that SSET encloses only atoms forming cycle-encoding superatoms, while RSET contains all others. As shown in Figure 3, there are many ways to generate these partitions. To expound the algorithm more clearly we shall trace the development of the second partition step by step. The argument to follow is obviously applicable to all other partitions.

The chosen partition has SSET = C_8OU_2 and RSET = C_1. A valence list (VL) is defined that contains the frequency of atoms of valence two, three, and four in the actual SSET. Because no atoms with valence three are present in the chosen example, we can write VL = (1,0,8). Valence one is not considered in a VL because there are no monovalent atoms in SSET.

It is now necessary to compute the free valences for the atoms in SSET. There are 14 free valences in total. The distribution of these free valences over the SSET atoms involves only the carbon atoms; the oxygen atom, being a member of SSET, cannot have any free valences left. One possible distribution pattern for the eight carbons is the following: (2,2,2,2,2,2,1,1,). (There is another one; can you name it?) Two carbon atoms each have one residual free valence; they are tertiary carbon atoms (and therefore must belong to the fused edge of two cycles). There are six further secondary carbon atoms. The superatom is a member of the class of two-trivalent-vertex graphs. This means that around the two vertex atoms two cycles of varying magnitude must be constructed, consisting of a total of six carbon atoms and one oxygen atom. As a double bond is considered a cycle, the candidate in Figure 3 containing a double bond and only one ring is a valid one, too.

After having generated all possible patterns for the two cycles by the secondary partition mechanism, each node in the resulting graph is attributed a chemical identity. At this stage, explicitly disposing of the free valences of the atoms, the RSET atoms (all monovalent) can be attached to the generated cyclic substructure. Here a wealth of possible permutational combinations are also feasible. However, once a candidate has reached its final structural completeness, the usual canonization algorithm is invoked immediately to check for redundant candidates. The final list of candidates is then ready for suitable output processing.

1. An Interactive Structure Generation Session

The following self-explanatory example reproduces a small portion of an original printout of a real investigation session run with CONGEN. Only a small fraction of the user-SES dialogue is reproduced here, but enough to demonstrate the interaction style between human and computer (text printed in capital letters is computer output).

WELCOME TO CONGEN VERSION VI.
CONGEN IS A PROGRAM FOR COMPUTER-ASSISTED STRUCTURE ELUCIDATION
DEVELOPED WITH NIH SUPPORT BY THE DENDRAL GROUP AT STANFORD.

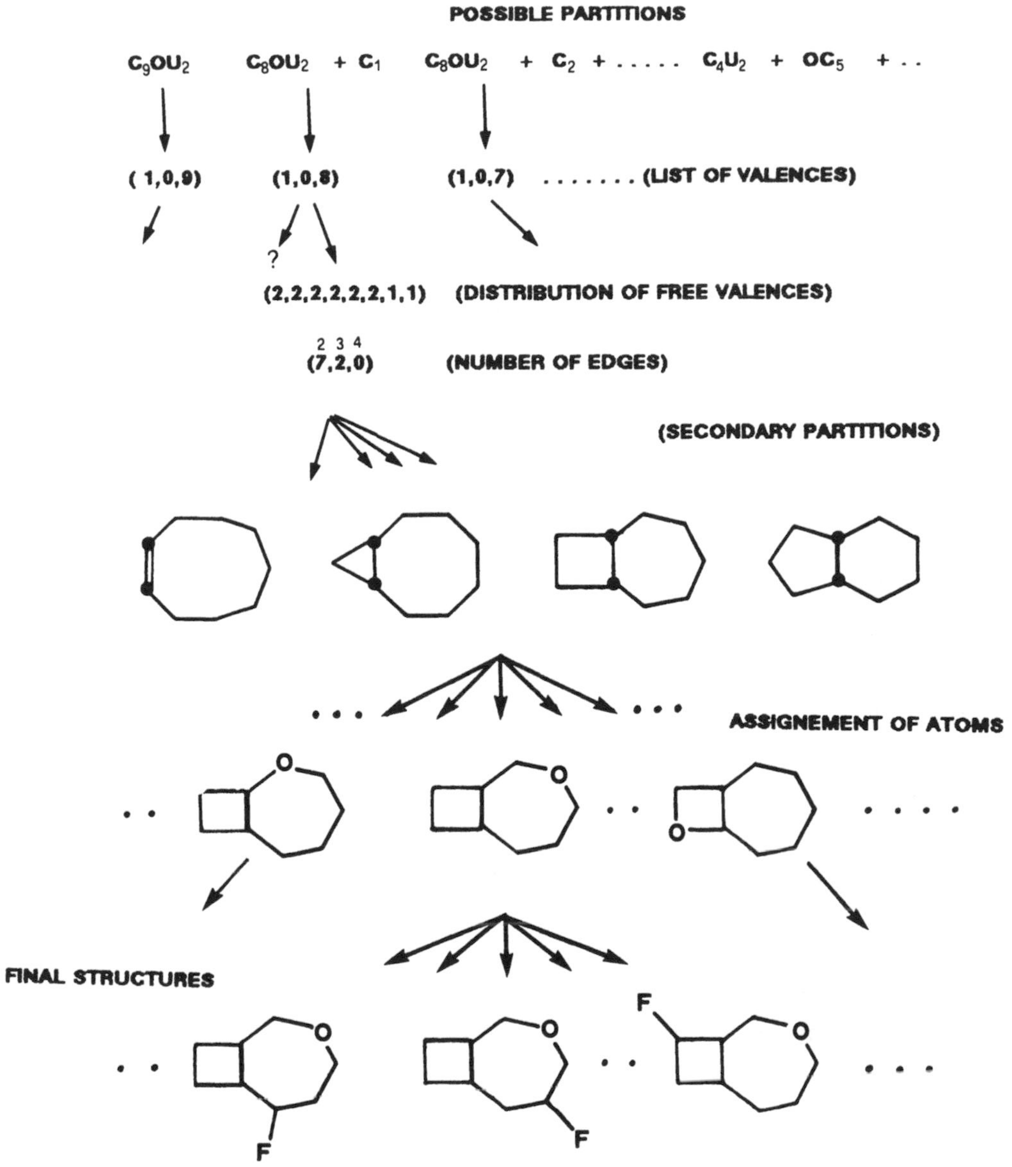

POSSIBLE PARTITIONS

C_9OU_2 C_8OU_2 + C_1 C_8OU_2 + C_2 + ... C_4U_2 + OC_5 + . .

(1,0,9)   (1,0,8)   (1,0,7)   . . . . **(LIST OF VALENCES)**

?

(2,2,2,2,2,2,1,1)   **(DISTRIBUTION OF FREE VALENCES)**

2 3 4
(7,2,0)   **(NUMBER OF EDGES)**

(SECONDARY PARTITIONS)

ASSIGNEMENT OF ATOMS

FINAL STRUCTURES

FIGURE 3.   Illustration of the generation of all possible isomers starting from a given SSET and RSET partition.

MAY I RECORD YOUR SESSION?: yes
PLEASE TYPE YOUR NAME: Hakkınen
@define molform c 9 h 12 o 2
@define sub keto
(NEW SUBSTRUCTURE)
>chain 2
>draw num

SUBSTRUCTURE KETO:

1-2

```
>join 1 2
>atname 2 c
>hrange 1 0 0
>draw atnamed
```

SUBSTRUCTURE KETO: (HRANGES NOT INDICATED)

C=O

```
>show
```

SUBSTRUCTURE KETO:

ATOM#	TYPE	NEIGHBORS	HRANGE
1	C	2 2	0—0
2	0	1 1	

```
>done
```
KETO DEFINED
@define sub methin
(NEW SUBSTRUCTURE)
```
>chain 1
>hrange 1 1 1
>show
```

SUBSTRUCTURE METHIN:

ATOM#	TYPE	NEIGHBORS	HRANGE
1	C		1—1

```
>done
```
METHIN DEFINED
@define sub methylene
(NEW SUBSTRUCTURE)
```
>chain 1
>hrange 1 2 2
>show
```

SUBSTRUCTURE METHYLENE

ATOM#	TYPE	NEIGHBORS	HRANGE
1	C		2—2

```
>done
```
METHYLENE DEFINED
@define sub cbone
(NEW SUBSTRUCTURE)
```
>ring 4
>branch 1 1
>join 1 5
>draw numbered
```

SUBSTRUCTURE CBONE:

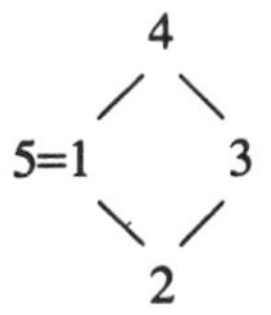

```
>atname 5 o
>hrange 2 0 2   3 0 2   4 0 2
>show
```

SUBSTRUCTURE CBONE:

ATOM#	TYPE	NEIGHBORS	HRANGE
1	C	5 5 4 2	
2	C	3 1	0—2
3	C	4 2	0—2
4	C	1 3	0—2
5	O	1 1	

```
>done
CBONE DEFINED
@generate
SUPERATOM:
'COLLAPSED FORMULA IS O 2 C 9 H 12
CONSTRAINT: sub methin exactly 2
CONSTRAINT: sub methylene exactly 5
CONSTRAINT: sub keto exactly 2
CONSTRAINT: sub cbone at least 1
CONSTRAINT:
......................................
```

```
41 STRUCTURES WERE GENERATED
@draw atnamed 10
```

#10:

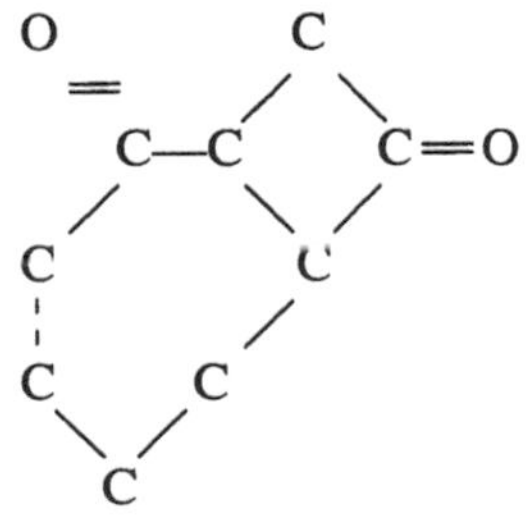

O   C
  =   / \
    C—C   C=O
   /   \ /
  C   C
  ¦   /
  C   C
   \   /
     C

```
@draw atnamed 20
```

#20:

```
        O
        ‖
      C—C—C—C
     /       \   \
    C         C—C
   /             ‖
  C              O
  ‖
  C
```

@draw atnamed 40

#40:

```
O
‖
  C --- C—C—C—C
  ⋮    /        \
  C—C            C
  ‖              ‖
O               C
```

@define sub ethylene
(NEW SUBSTRUCTURE)
>chain 2
>join 1 2
>done
ETHYLENE DEFINED
@survey

DO YOU WANT TO USE A LIBRARY OF SUBSTRUCTURES? no
SUBSTRUCTURE NAME: ethylene
SUBSTRUCTURE NAME:

SCANNING THROUGH STRUCTURES.

....

@STRUCTURES WITH DISCRIMINATING FEATURES:
     10   ETHYLENE
DO YOU WANT TO SELECT STRUCTURES WITH COMBINATIONS OF FEATURES?
no
@prune
CONSTRAINT: sub ethylene none
CONSTRAINT:
.....**....*.......*..***.....*...*....*.
31 STRUCTURES SURVIVED PRUNING

The session continues in a similar manner with the definition of other positive and
negative constraints which lead to the pruning of many other candidates, until a very small
number survive. The pedagogical essence of this example session is that all substructures,
like methin, cbone, etc., must be given by the user to the SES. In practice, this forces the

chemist to interpret chemical and spectral data by himself in order to gain reliable information about the present constraints. Here the PLANNING phase is human dependent.

This example also nicely illustrates how direct visual inspection of a few sample candidates (#10, #20, and #40) provokes in the human mind the sudden perception of some substructures that cannot be present in the unknown compound. Suspecting or even sensing the impossiblity of their physical existence is a consequence of the chemist's sudden awareness of their predicted theoretical existence. This awareness is brought about in two ways: by the visual inspection of the printout drawings and by the previous knowledge of the chemist, who has analyzed the spectra on his own. The resulting effect is an induced reinterpretation of the available experimental data that will confirm the negative constraints. The visual perception of structure #40 is the primer to the subsequent pruning of any candidate containing an ethylene substructure, evidently not in accordance with the available spectral and/or chemical data.

As already mentioned, SES systems with automatic constraint selection (done principally in the PLANNING phase) have been developed. The following section will present two such important systems, CASE and CHEMICS, underlining their superior performance coming from a fully automated, autodeductive PLANNING phase.

III. SES WITH AUTODEDUCTIVE INTERPRETATION OF SPECTRAL DATA

The SES programs CASE[6,7] and CHEMICS[8-11] are the result of many man-years of effort in the design of fully automatic, self-deciding structure elucidation systems. They have powerful modules responsible for the transformation of raw spectral data into positive constraints. CHEMICS accepts the following data:

- The molecular formula
- Low-resolution mass spectra for determining the molecular formula
- ^{1}H-NMR data for the selection of suitable fragments, called components, present in the unknown
- ^{13}C-NMR data for the same purpose
- IR data for the same purpose
- MACRO input: user-defined positive and negative constraints
- 2D-NMR data

A component is a small substructural unit consisting of one centered atom surrounded by a shell of neighbors. The component must have at least one free valence left. The principal task of the spectrum interpreter in CHEMICS is to correlate the spectral data to one or more such components, which are contained in a limited number in a small, program-internal library. Inside the library each component is linked to a particular shift range for NMR data and to a specific wave number range for IR data.

An IR-spectrum interpreter has recently been developed for CHEMICS on the basis of symbolic logic in order to formulate complex relationships between the substructures and their characteristic wave number regions.[11]

A common methodology for the correlation of ^{1}H- and ^{13}C-NMR signals is currently used in CHEMICS; the contribution made by the carbon NMR interpreter will be discussed in more detail.

A. THE ^{13}C-NMR INTERPRETER

ASSINC is the module responsible for substructural inference from autodeductive interpretation of ^{13}C-NMR chemical shifts, spin coupling, and signal intensity. The program is subdivided into four parts: (1) a procedure accepting the input; (2) a set of procedures for the primary analysis, which consists of attributing a specific type of carbon atom to each signal; (3) a procedure for the secondary analysis, which verifies that the components selected in the primary analysis are not globally contradicting the experimental spectrum; and, finally, (4) a library of chemical shifts. The scheme below represents the architecture of the PLANNING phase of CHEMICS:

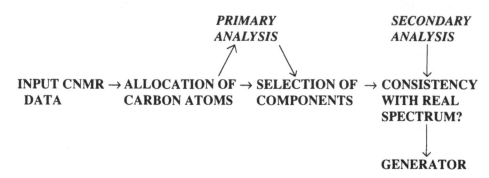

The standard version of CHEMICS contains 189 components in the internal file. A short excerpt of this file is listed here. A newer version of CHEMICS uses components according to a revised component hierarchy that involves 12 basic attributes to determine the bonding priorities of the different components.

#	COMPONENT		SHIFT RANGE (ppm)
.	.		.
.	.		.
.	.		.
18	CH₃–(C)–	(C)	9.9 **** 12.9
19	ISOPROPYL	(O)	15.0 **** 25.8
20	ISOPROPYL	(A)	16.6 **** 25.8
21	ISOPROPYL	(Y)	20.9 **** 25.4
22	ISOPROPYL	(K)	15.0 **** 23.8
23	ISOPROPYL	(D)	16.3 **** 25.8
24	ISOPROPYL	(T)	15.0 **** 25.8
25	ISOPROPYL	(C)	15.2 **** 25.5
26	CH₃O–	(O)	52.8 **** 61.6
27	CH₃O–	(Y)	54.5 **** 57.9
28	CH₃O–	(K)	50.3 **** 52.5
29	CH₃O–	(D)	56.6 **** 61.5
30	CH₃O–	(T)	52.8 **** 61.5
31	CH₃O–	(C)	49.9 **** 60.6
32	CH₃O–	(Y)	7.2 **** 26.1
.	.		.
.	.		.
.	.		.

Identical components are characterized by the changing adjacent molecular environment, which causes the diversification of the shift ranges. The environment to which a component could be attached is coded with labels: O, Y, K, D, T, and C are labels for oxygen, aromatic carbon, carbonylic carbon, olefinic carbon, acetylenic carbon, and saturated carbon, respectively.

As soon as the raw spectral data are input, the interpreter starts with the determination of the number of carbon atoms corresponding to each spectral peak (or peak pattern, if signal splitting occurs when C–H coupling is not suppressed).

Unlike in hydrogen NMR (HNMR) spectroscopy, the integral of carbon NMR (CNMR) peaks is not always linearly related to the number of resonating nuclei. ASSINC, however, assumes this linear dependence in a first approximation for carbon nuclei carrying hydrogen atoms due to the progressive action of the nuclear Overhauser effect. The algorithm uses the signal multiplicity data to partition the set of carbon atoms into two groups: those having at least one hydrogen as a bond partner and those having none. Through an empirical equation the number of atoms responsible for a certain resonance pattern (the carbon allocation number) is calculated. With the number of carbon atoms and their individual degree of hydrogenation now known, the system initiates the selection of possible components.

The presence or absence of a certain component is determined from the degree of hydrogenation and from the chemical shift of the allocated carbon atom, a shift that can vary within a certain range. Every component has its own range within which a signal must be located in order to activate that particular component.

From the number of signals (N) and the number of components (M) extracted in the primary analysis, a so-called NM matrix is generated. NM matrix elements are the carbon allocation number (e_{mn}) for the individual components that show shift ranges compatible with the peak shifts.

The NM matrix has the following structure:

SIGNAL	1	2	3	4	5	6	.	.	.	.	.	.	.	N
COMPONENT 1	e_{11}	.	.	.	.	.	.	.	.	.	.	.	.	e_{1N}
COMPONENT 2	.	e_{22}												
COMPONENT 3	.	.		.										
.	.	.												
.	.	.												
.	.	.												
COMPONENT M	e_{M1}	.	.	.	.	.	.	.	.	.	.	.	.	e_{MN}

In the secondary analysis the components selected from the internal file are grouped into a maximum number of different component sets (CS), which are in accordance with the molecular formula. This last condition is certainly not sufficient to consider a CS eligible for the subsequent GENERATING phase. Consistency with the real spectrum must be ensured first. This is achieved by a manipulation of the NM matrix, which is rewritten substituting variables (x) for its nonzero elements e_{nm}. For example, a hypothetical NM matrix shall contain 36 components extracted from the primary analysis, related to, say, four peaks. Suppose that six different CSs can be generated that are coherent with the molecular formula, yielding six different matrices $X(4,4)$. To further illustrate the example, assume the first component in any one of the available X to be compatible with peak 1 as well as with peak 2. Component 2 is attributed to peak 3; components 3 and 4 are both judged to be potentially responsible for peak 4. The matrix X, then, has the following form:

PEAK	1	2	3	4
COMPONENT 1	x_1	x_2	0	0
COMPONENT 2	0	0	x_3	0
COMPONENT 3	0	0	0	x_4
COMPONENT 4	0	0	0	x_4

The sum of j elements over a row must be equal to the number of carbon atoms c_i of the component i,

$$\sum_j x_{ij} = c_i \quad ; \quad \mathbf{c} = (c_1, c_2, \ldots, c_m)$$

and the sum over a column must result in the number of resonating carbon nuclei r for each signal j (i.e., the peak integral):

$$\sum_i x_{ij} = r_j \quad ; \quad \mathbf{r} = (r_1, r_2, ..., r_n)$$

Matrix $\mathbf{X}$, together with the vectors $\mathbf{c}$ and $\mathbf{r}$, generates a pair of simultaneous linear equations,

$$\mathbf{X\,I} = \mathbf{c} \quad \text{(column vector)} \tag{1}$$

$$\mathbf{I\,X} = \mathbf{r} \quad \text{(row vector)} \tag{2}$$

where $\mathbf{I}$ refers to an identity vector. If the above equation systems have an unambiguous solution the corresponding CS is saved; otherwise it is eliminated. Nevertheless, a CS validated in the secondary analysis can lead to incorrect candidates in the GENERATING step. To understand the reason for this an oversimplified example is constructed. Let a hydrogen-decoupled CNMR spectrum have just two peaks, with an intensity ratio of 2:1 in favor of peak 1, which is located at a higher field. Also let the primary analysis select the following types of components according to the shift ranges:

- Aliphatic carbon atoms linked to another carbon atom C–(C)
- A carbon atom linked to oxygen, C–(O)

Let the molecular formula be C_3H_8O. The possible component sets are

- CS1: {HO–CH< (isopropyl),CH_3–, CH_3–} → C_3H_8O
- CS2: {CH_3CH_2–,HO–CH_2–} → C_3H_8O

For CS1 we have for the equation systems

		Peak no.		
Component no.		1	2	c
(HO–CH<)	1	0	x_2	= 1
(2CH$_3$)	2	x_1	0	= 2

and, transposing $\mathbf{r}$ for more clarity,

	Component no.		
Peak no.	1	2	r
1	0	x_1	= 2
2	x_2	0	= 1

The solutions for both systems are $x_1 = 2$ and $x_2 = 1$. CS1 is a valid set.
    Taking CS2 one obtains

		Peak no.		
Component no.		1	2	c
(HO–CH$_2$–)	1	0	x_2	= 1
(CH$_3$CH$_2$–)	2	x_1	0	= 2

and

	Component no.		
Peak no.	1	2	r
1	0	x_1	= 2
2	x_2	0	= 1

These combined equation systems have the same solutions as the previous ones. Also, CS2 has to be considered valid and can be passed on to the structure generator, but the only possible structure obtainable from CS2 is propanol, which has three topologically and magnetically different carbon atoms. However, we decided to build the example on a spectrum having only two signals. CS1 offers isopropanol as a unique solution; indeed, it has only two different types of resonating carbon atoms because the two methyl groups are magnetically (and canonically) equivalent. It follows that some additional evaluation must be appended to the SES after the generation of the candidates is terminated. This check belongs to the TESTING phase. (The above example is purposely oversimplified to focus attention on the latent ambiguity of the secondary analysis; in reality, the C–H spin coupling pattern would at once reveal CS1 to be the only possible set.)

B. TESTING WITH CHEMICS

In many (but not in all) cases, topologically equivalent carbon atoms are also magnetically equivalent. The CHEMICS TESTING approach is realized by a predictor of the number of peaks expected in the CNMR spectrum of each candidate. This number is compared with the real number of peaks of the unknown molecule, and all candidates not in line with the experimental evidence are deleted. The predicted number of peaks is calculated from the number of topologically different classes of carbon atoms, information easily obtained from the application of canonical numbering algorithms to the structures of the candidates.

Two complete CHEMICS demonstration runs are shown below as vivid examples of the performance of this SES, a program so important in the history of computer chemistry research. In both cases the SES does not deliver a unique solution, but this is seldom the case. Nevertheless, it is of great importance to the chemist that the range of possibilities becomes drastically restricted. The final identification of the true compound among the few predicted by the computer is definitely not a serious problem for the experimentalist at this point.

CHEMICS example 1

4,6,6-Trimethylbicyclo[3.1.1.]hept-3-en-2-one
Verbenone

INPUT DATA

DATA
C10 H14 01

SPECTRAL DATA

H-NMR			IR			C-NMR			
NO.	POSI.	AREA	NO.	POSI.	INT.	NO.	POSI.	HEIGHT	MULT.
1	339.00	10.00	1	2940	79	1	21.90	139.00	4
2	337.00	23.00	2	1777	31	2	23.30	882.00	4
3	333.00	18.00	3	1720	45	3	26.40	937.00	4

H-NMR			IR			C-NMR			
NO.	POSI.	AREA	NO.	POSI.	INT.	NO.	POSI.	HEIGHT	MULT.
4	331.00	4.00	4	1678	97	4	40.60	950.00	3
5	185.00	5.00	5	1620	70	5	49.60	805.00	2
6	178.00	6.00	6	1475	44	6	53.60	308.00	1
7	173.00	12.00	7	1436	62	7	57.50	882.00	2
8	168.00	13.00	8	1376	63	8	121.00	779.00	2
9	163.00	12.00	9	1338	65	9	169.70	377.00	1
10	158.00	25.00	10	1280	51	10	203.00	212.00	1
11	153.00	24.00	11	1239	62				
12	149.00	18.00	12	1228	33				
13	145.00	24.00	13	1200	60				
14	141.00	6.00	14	1078	33				
15	136.00	5.00	15	1030	57				
16	124.00	50.00	16	977	35				
17	120.00	146.00	17	856	52				
18	119.00	100.00	18	755	37				
19	115.00	26.00							
20	89.00	190.00							
21	59.00	189.00							

INPUT MACRO

DATA
C=CH–CO

OBTAINED STRUCTURES

STRUCTURE NO. 1

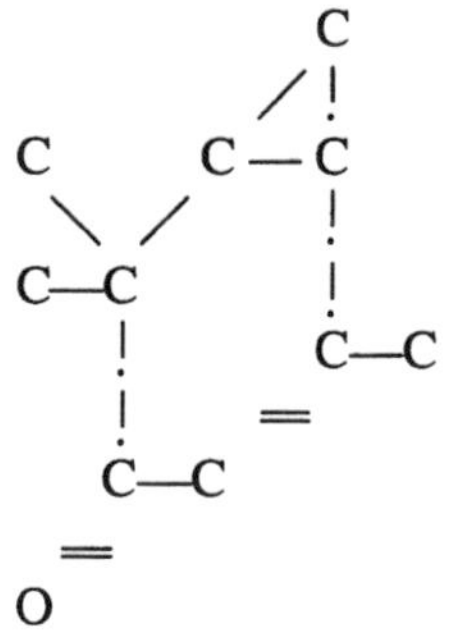

STRUCTURE NO. 2

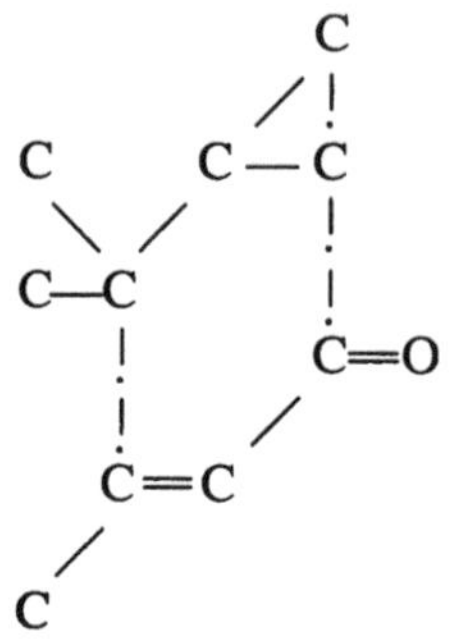

STRUCTURE NO. 3

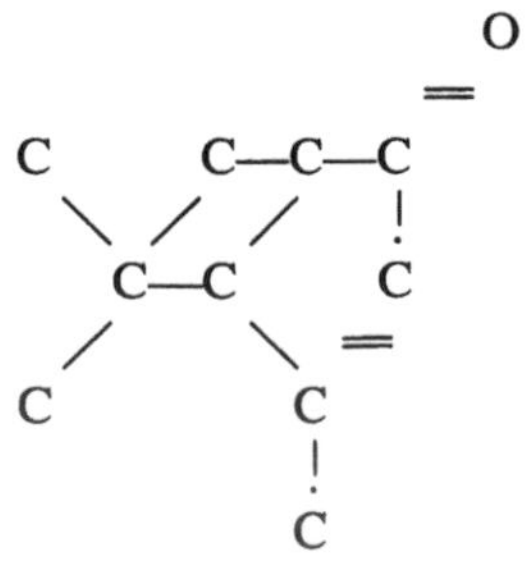

STRUCTURE NO. 4

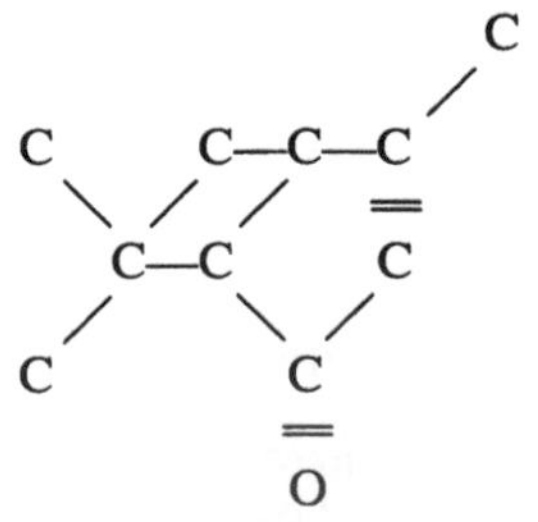

STRUCTURE NO. 5   STRUCTURE NO. 6

STRUCTURE NO. 7

CHEMICS example 2

1,3,3-Trimethyl-2-oxabicyclo[2.2.2]octane
Cineole

INPUT DATA
C10 H18 O1

SPECTRAL DATA

	IR			C-NMR FOLLOWING C-NMR DATA WERE ACCEPTED		
NO.	POSI.	INT.	NO.	POSI.	HEIGHT	MULT.
1	2950	93				
2	2900	92	1	22.80	830.00	3
3	1450	67	2	27.50	400.00	4
4	1380	76	3	28.80	930.00	4
5	1250	81	4	31.50	880.00	3
6	1080	82	5	32.90	330.00	2
7	1000	92	6	69.60	140.00	1
8	860	64	7	73.50	145.00	1

OBTAINED STRUCTURES

STRUCTURE NO. 1

```
        C
        |\
    C   | \  C
    |   |  \ / \
C--C--O--C     C
    |      \   |
    C       C —C
```

STRUCTURE NO. 2

```
            C
           / \
    C     C ---C
    |    /
C--C--O--C
    |    \
    C     C -- C
```

STRUCTURE NO. 3

```
    C       C—C
    |       |  |
C—C- -O--C—C
    |       |  |
    C       C—C
```

STRUCTURE NO. 4

```
              C
              |\
    C         | \  C
    |         |  \ / \
C--C--O--C—C—C
    |         |  \ / |
    C         C
```

STRUCTURE NO. 5

```
  C    O
   \  / \
    C --- C--C—C—C
   /      |    \ /
  C       C    C—C
```

STRUCTURE NO. 6

```
  C        C         C
   \      |         / \
    C --- C--C--C     C
   /      |         \ /
  C       O          C
```

STRUCTURE NO. 7

```
        C
        |
  C     |
   \    |
    C—C- -C- -C- -C—C
   /    |              \ |
  C     O               C
```

STRUCTURE NO. 8

```
  C          C—C
   \         |  |
    C—C --- C     C
           |  \  /
           O—C—C
               |
               C
```

STRUCTURE NO. 9

```
        C       C
        |      / \
  C     O—C- -C     C
   \   /         \ /
    C—C           C
   /
  C
```

STRUCTURE NO. 10

```
          C—C
         /    \
    C         C
    |          |
C—C            C—C
   \          /
    C        C
     \      /
       O
```

STRUCTURE NO. 11

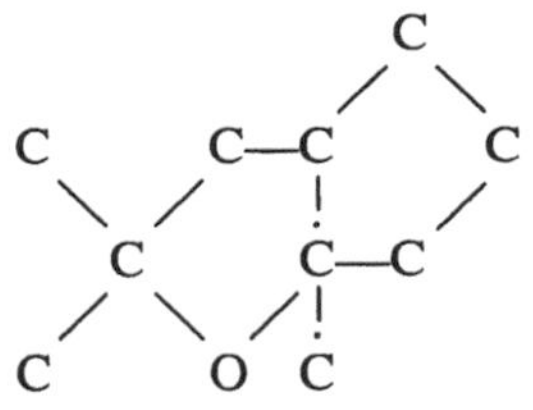

STRUCTURE NO. 12

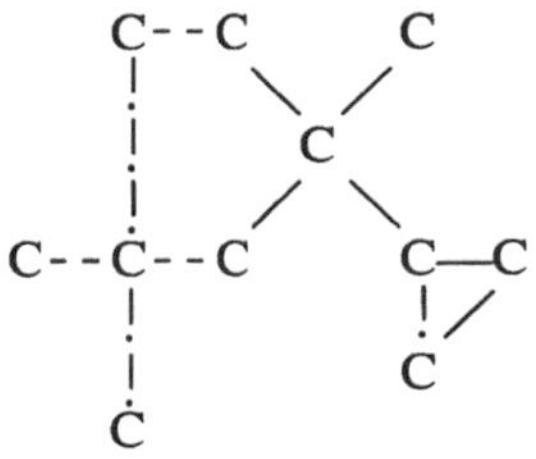

STRUCTURE NO. 13

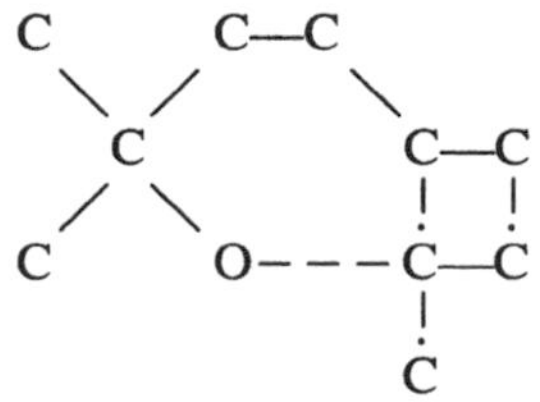

STRUCTURE NO. 14

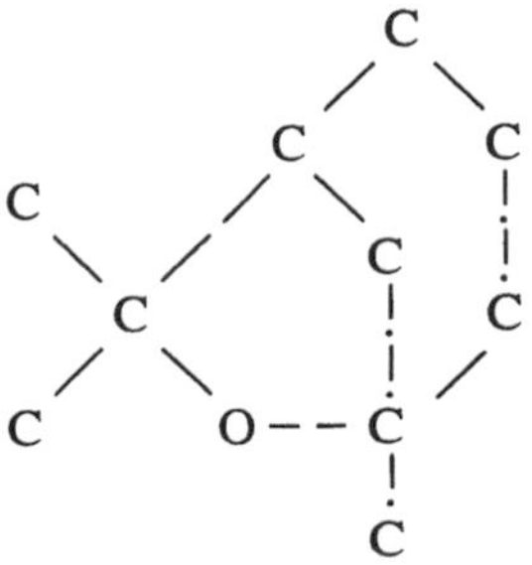

STRUCTURE NO. 15

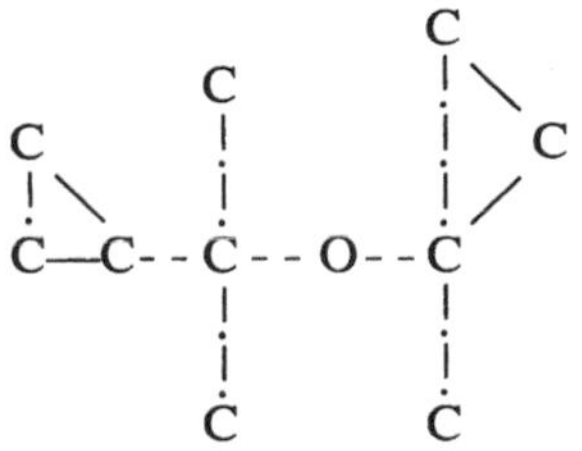

STRUCTURE NO. 16

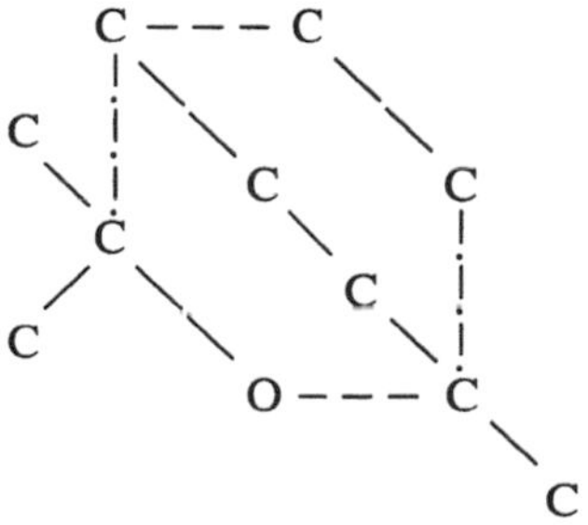

STRUCTURE NO. 17

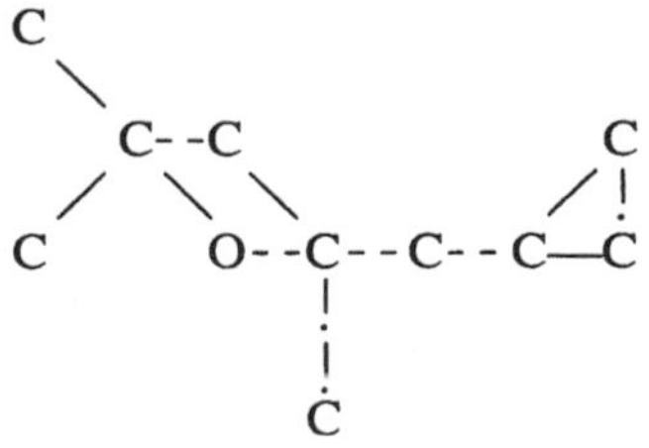

STRUCTURE NO. 18

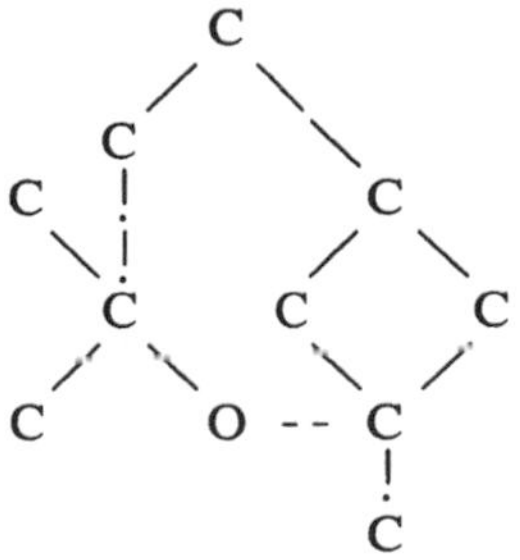

STRUCTURE NO. 19

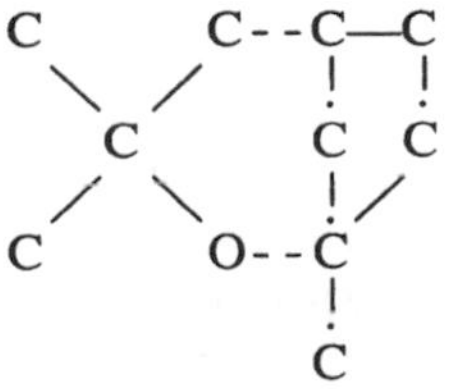

C-13 NMR SIGNAL NUMBER PREDICTION

** TOTAL NUMBER OF STRUCTURES   19

PEAK PREDICTION RESULTS
OBSERVED SIGNAL NUMBER   7

PREDICTED   NO. OF STRUCTURES
SIGNAL NO.
7   3

** TOTAL NO. OF HITTING STRUCTURES   3

$$$ END OF PREDICTION $$$

OBTAINED STRUCTURES

STRUCTURE NO. 1   STRUCTURE NO. 2   STRUCTURE NO. 3

IV. THE CASE SYSTEM

CASE is an SES which shows some affinity to CHEMICS. It is geared around a substructural inference machine exploiting spectral data. It accepts as input IR, HNMR, ^{13}C-NMR, and 2-D NMR data, as well as user-given substructural information.

The information block (molecular formula and spectral data) is processed by the INTERPRET module, which generates a list of substructural fragments, the ACFs, compatible with the input. The prototype version of CASE has a file of 5088 "basic units of structure" (ACFs).[6,7]

Two inference machines have been newly integrated into the system. INFERCNMR and INFER2D are both responsible for predicting carbon substructural skeletons. An infrared interpreter has been developed as well. CASE differs from CHEMICS in the evaluation of the CNMR data. The latter program uses a file of components already connected to some shift ranges, while the former SES contains a mighty library of complete ^{13}C-NMR spectra (currently about 10,000) of reference compounds. INFERCNMR is essentially a subspectrum matching procedure that explicitly retrieves atom substructures with their hybridization states, degrees of hydrogenation, bond types, and heteroatomic partners. The input consists of shift and multiplicity for each signal of the experimental spectrum.

INFER2D processes 2-D spectroscopic data: three-bond proton-proton correlations (COSY experiment), one-bond proton-carbon correlations, long-range proton-carbon correlations, and the results of the 2D INADEQUATE experiment.

The most valuable information from a strategic aspect is the carbon-carbon signal con-

TABLE 1
The Input Data for the CASE Elucidation Run of Monochaetin

Monochaetin formula: $C_{18}H_{20}O_5$

	^{13}C-NMR data			Proton NMR data		
	Shift	Multiplicity		Shift	Integral	Coupling
1.	205.94	S	1.	6.79	1	
2.	191.77	S	2.	6.02	1	
3.	169.10	S	3.	5.29	1	
4.	158.52	S	4.	4.05	1	
5.	145.52	S	5.	3.76	1	
6.	143.30	D	6.	3.19	1	
7.	116.22	S	7.	2.13	3	D J = 0.4 Hz
8.	107.04	D	8.	1.81	1	
9.	105.73	D	9.	1.48	1	
10.	82.55	S	10.	1.32	3	S
11.	52.13	D	11.	1.11	3	D J = 6.7 Hz
12.	46.70	D	12.	0.97	3	T J = 7.4 Hz
13.	43.66	D	Total		20	
14.	26.27	T				
15.	19.49	Q				
16.	18.92	Q				
17.	14.39	Q				
18.	11.45	Q				

nectivity. This information can be derived directly from 2D INADEQUATE experiments. The INFER2D module transforms the COSY data and the proton-carbon correlations into carbon-carbon signal connectivity: if molecular symmetry is absent, this is equivalent to carbon-carbon atom connectivity because of the one-to-one relation between signal and carbon atom. However, the postulated substructures may not always be assorted unambiguously with regard to carbon hybridization, bond type, and neighboring heteroatoms. When symmetry occurs, special algorithms use group theory in attempting to manage the multiple possibilities of signal-to-atom correlations. Different assignment patterns give rise to different molecular symmetries. Thus, a number of substructural interpretations must be kept open until more evidence (e.g., overlapping substructures) becomes available.

The collected substructural units, the ACFs, which have survived the necessary spectrum coherence check unpruned, are conveyed to the COCOA (Constrained Combination of ACFs) structure generator. The novel working strategy of COCOA, hyperstructure reduction, has been discussed previously.

In addition to the usual substructural constraints, other tests refine the structure reduction process, improving its performance and speed. For example, chemically "impossible" strained compounds are deleted (e.g., strained bridgehead double bonds).

A. AN EXAMPLE WITH CASE: MONOCHAETIN

The following example should illustrate the performance of CASE. Monochaetin is a fungal metabolite of molecular formula $C_{18}H_{20}O_6$. The reported structure assignment[11] depended heavily on 2-D NMR experiments — in particular, long-range proton-carbon correlations. With the capability of INFER2D to interpret such data, the problem has been of special interest to the CASE team. In Table 1 the raw data for CNMR and HNMR experiments are given. A summary of the user-entered correlations observed in the 2-D NMR measurements is presented in Table 2.

The input consists of the chemical shifts of the signals that have been correlated, the

TABLE 2
The User-Entered Correlations Observed in Two-Dimensional NMR Experiments on Monochaetin

	Shift 1	Shift 2	Min.	Max.
	143.30 (C)	6.79 (H)	1	1
	107.04 (C)	6.02 (H)	1	1
	105.73 (C)	5.29 (H)	1	1
	43.66 (C)	3.76 (H)	1	1
	52.13 (C)	4.05 (H)	1	1
C-H correlation	46.70 (C)	3.19 (H)	1	1
	26.27 (C)	1.81 (H)	1	1
	26.27 (C)	1.48 (H)	1	1
	11.45 (C)	0.97 (H)	1	1
	19.49 (C)	2.13 (H)	1	1
	18.92 (C)	1.32 (H)	1	1
	14.39 (C)	1.11 (H)	1	1
	3.19 (H)	1.11 (H)	3	3
	3.19 (H)	1.81 (H)	3	3
COSY	3.19 (H)	1.48 (H)	3	3
	0.97 (H)	1.48 (H)	3	3
	0.97 (H)	1.81 (H)	3	3
	46.70 (C)	1.11 (H)	2	3
	205.94 (C)	1.11 (H)	2	3
	26.27 (C)	1.11 (H)	2	3
	82.55 (C)	1.32 (H)	2	3
	43.66 (C)	1.32 (H)	2	3
	191.77 (C)	1.32 (H)	2	3
	82.55 (C)	3.76 (H)	2	3
	52.13 (C)	3.76 (H)	2	3
	205.94 (C)	3.76 (H)	2	3
	143.30 (C)	3.76 (H)	2	3
	116.22 (C)	3.76 (H)	2	3
	169.10 (C)	4.05 (H)	2	3
	43.66 (C)	4.05 (H)	2	3
Long-range	205.94 (C)	4.05 (H)	2	3
C-H correlation	82.55 (C)	5.29 (H)	2	3
	116.22 (C)	5.29 (H)	2	3
	107.04 (C)	5.29 (H)	2	3
	105.73 (C)	6.02 (H)	2	3
	145.52 (C)	6.02 (H)	2	3
	116.20 (C)	6.02 (H)	2	3
	158.52 (C)	6.02 (H)	2	3
	19.49 (C)	6.02 (H)	2	3
	116.20 (C)	6.79 (H)	2	3
	145.52 (C)	6.79 (H)	2	3
	158.52 (C)	6.79 (H)	2	3
	3.76 (H)	4.05 (H)	3	3
	158.52 (C)	2.13 (H)	2	3
	107.04 (C)	2.13 (H)	2	3

element type, and the minimum and maximum number of bonds between the atoms corresponding to these signals. One-bond proton-carbon correlations have been entered first, three-bond proton-proton correlations next, and, finally, all long-range proton-carbon correlations that do not distinguish between two and three intervening bonds. The substructures inferred by INFER2D are listed in Table 3.

TABLE 3
List of Substructures Inferred by INFER2D

Substructure Constraints from Two-Dimensional NMR

C52.13 .C43.66;C46.7 (.C14.39).C26.27 .C11.45

C205.94 C14.39	C205.94 A C14.39
C26.27 C14.39	C26.27 A C14.39
C82.55 C18.92	C82.55 A C18.92
C43.66 C18.92	C43.66 A C18.92
C191.77 C18.92	C191.77 A C18.92
C82.55 C43.66	C82.55 A C43.66
C205.94 C43.66	C205.94 A C43.66
C143.3 C43.66	C143.3 A C43.66
C116.22 C43.66	C116.22 A C43.66
C169.1 C52.13	C169.1 A C52.13
C205.94 C52.13	C205.94 A C52.13
C82.55 C105.73	C82.55 A C105.73
C116.22 C105.73	C116.22 A C105.73
C107.04 C105.73	C107.04 A C105.73
C105.73 C107.04	C105.73 A C107.04
C145.52 C107.04	C145.52 A C107.04
C116.22 C107.04	C116.22 A C107.04
C158.52 C107.04	C158.52 A C107.04
C19.49 C107.04	C19.49 A C107.04
C116.22 C143.3	C116.22 A C143.3
C145.52 C143.3	C145.52 A C143.3
C158.52 C143.3	C158.52 A C143.3
C158.52 C19.49	C158.52 A C19.49

First, two different carbon atom substructures (a two-carbon and a four-carbon fragment, separated by a semicolon) are predicted to be present. The specific constituent carbon atoms of these required substructures are designated in terms of their chemical CNMR shifts. Each remaining line lists two alternative representations of the long-range proton-carbon correlations. The substructure on the left assumes two-bond coupling; on the right, three-bond coupling. In the three-bond coupled sequence the central element A can be any nonhydrogen element. The real work of interpreting the full structural significance of these substructures remains to be done by COCOA.

Figure 4 contains five user-entered positive constraints. This substructural information, together with that derived from INFER2D, is the basis for COCOA's "intelligent" work. Only one final structure is generated which shows the correct result: monochaetin (Figure 5).

B. ANOTHER EXAMPLE: FORGETTING STRUCTURES

With an increasing number of substructures it becomes difficult for the human mind to construct all possible assembly combinations. Structures may then be "forgotten" easily due to the limited combinatoric power of the human mind. This is a deficiency not encountered when working with an SES.

^{13}C-NMR spectroscopy was used in an attempt to elucidate the structure of the product obtained from thermolysis of a compound shown in Figure 6.[13] Six different signals were measured with the following parameters: (1) 210.3 ppm, singlet; (2) 130.8 ppm, doublet; (3) 47.8 ppm, singlet; (4) 35.6 ppm, triplet; (5) 24.7 ppm, quartet; and (6) 20.1 ppm, triplet.

Five different types of substructural units were inferred manually, shown here with their frequencies:

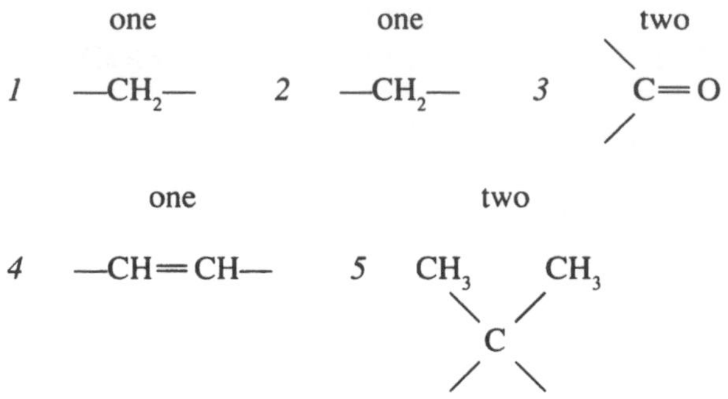

Four different final structures were totally generated from the above fragments as shown in Figure 7.

The same problem was processed with CASE, and 108 candidates were generated. Using the command PEAK a prediction of the number of peaks in the ^{13}C-NMR spectrum of each candidate was obtained. Only six candidates survived the test, having the required symmetry properties. In addition to the postulated four candidates, two more were generated autodeductively by the SES, solutions which were forgotten in the manual elucidation process.

STRUCTURE 106   STRUCTURE 52

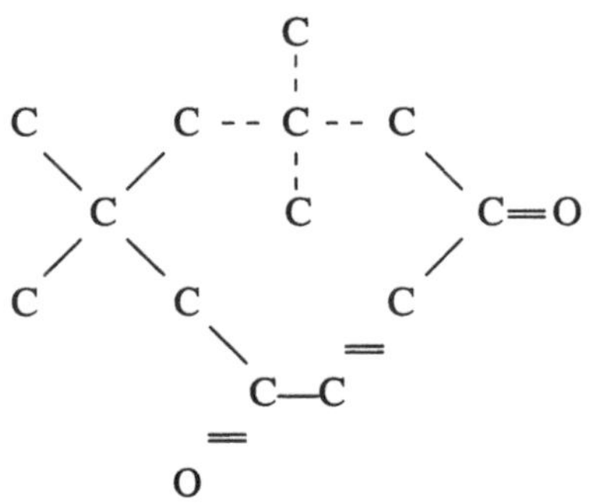

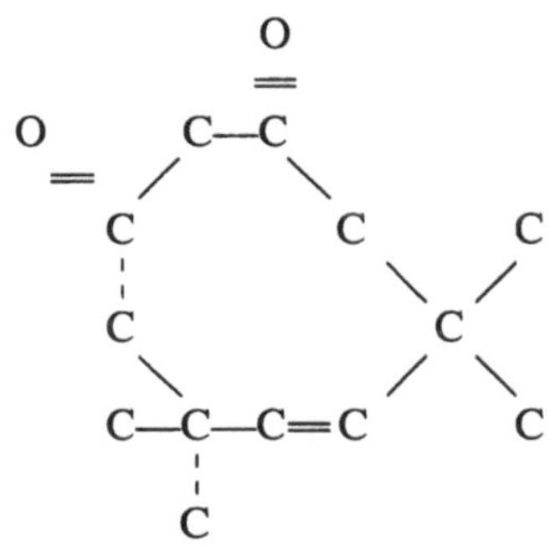

V. TESTING USING MASS SPECTRA PREDICTORS

The DENDRAL SES project has been extremely active in the development of structure elucidation methodologies in a variety of fields.

Much work has been done in the area of TESTING modules. Some of these modules in particular deserve separate presentations. One, PREDICTOR, is an AI-oriented software system that simulates mass spectra of molecules. If these molecules are the candidates coming from the GENERATING step, it can be seen as a further testing tactic of an SES. PREDICTOR[14] is based on two separate formal concepts: the Half-Order Theory (HOT) and the Rule-Based Theory (RBT).

A. THE HALF-ORDER THEORY AND MASS SPECTROSCOPY SIMULATION

This theory received its name because it tries to model processes occurring in mass spectrometry in such a simple approximation that it does not even reach the level of "first approximation" in the opinion of the DENDRAL team. In HOT the following points are assumed:

1.   Every chemical bond in a molecule can, in principle, be fragmented in an MS process if not subjected to special constraints.

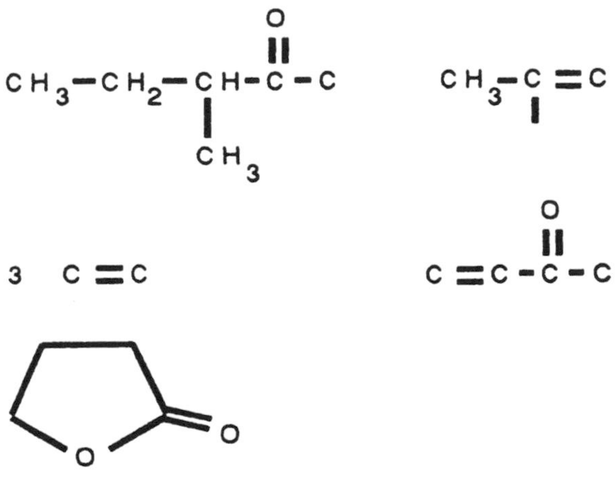

O
‖
$CH_3-CH_2-CH-C-C$ $CH_3-C=C$
| |
CH_3

3   $C=C$ O
                                 ‖
                      $C=C-C-C$

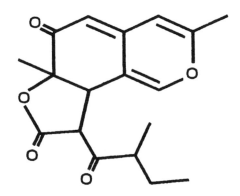

FIGURE 4.   Five user-entered positive constraints.

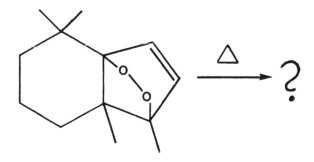

FIGURE 5.   The correct structure of monochaetin.

FIGURE 6.   This compound is decomposed thermally, yielding an unknown product which is the object of a structure elucidation study with CASE.

2.   Inside the fragmentation algorithm there are predefined constraints forbidding "impossible" bond ruptures from an MS point of view.
3.   A coefficient of probability is attributed to each simulated peak of the predicted MS spectrum.
4.   Typical constraints present in HOT include the following:

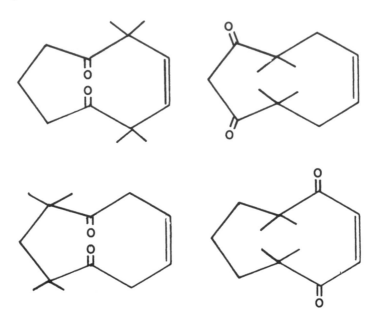

FIGURE 7.   The four structures already known for the unknown product, obtained from manually assembling the five substructures shown in the text.

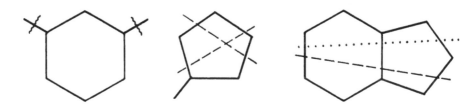

FIGURE 8.   Bond-breaking patterns for generating two fragments from a molecular structure in MS simulation programs.

A. Only one- and two-step fragmentations are possible.
B. During each step a maximum of two bonds can be broken.
C. Rupture of aromatic bonds or of isolated double and/or triple bonds is not allowed.
D. Simultaneous multiple ruptures of bonds between a central atom and nonhydrogen atoms are not allowed.
E. Multiple hydrogen interchanges between fragments are not allowed.

PREDICTOR is instructed to break three classes of bonds: the first class includes all acyclic bonds, the second class consists of pairs of bonds in rings, and the third class is composed of bond triples in condensated ring systems.

A special algorithm classifies the bonds, creating cutsets, by examining the topological structure of a molecule. A cutset pertaining to the first class can be generated easily, as it contains any bond that is not a member of a ring. The search for the elements of cutsets of the two other classes requires more effort. After selection of a first bond member of a ring, the algorithm must select a second one (a two-bond cutset) or a third one (a three-bond cutset); if broken along with the first bond, it gives rise to two molecular fragments consisting of at least one central carbon atom (in order not to collide with the fourth constraint issued above). Figure 8 gives some possible bond rupture patterns of simple structures as examples. Another part of the program generates all combinations of cutsets, leading to the simulated

MS ions. The user can control the simulation by means of adaptable parameters (e.g., the total number of bonds to be broken during a sequence of fragmentations). These bonds are processed stepwise through the repeated transformations of the molecular graph according to the cutsets sequentially chosen and applied by the system.

All of the peaks predicted so far have unitary intensity (equal probability). If he wishes, the user can define a degree of probability for bond breaking. It follows that peaks will show a simulated intensity, roughly reproducing the abundance of formation of a particular ionic fragment. The completed simulated spectrum is matched to the real spectrum to locate corresponding peaks. The excessive flexibility of HOT must account for the prediction of phantom peaks that are never measured in reality. They are eliminated manually. The accumulated experience, especially facing phantom peaks, permits the programmer to enhance the intelligence, the chemical brain, of the chemical software and to develop refined versions of the program prototypes. This feedback mechanism is common to all computer chemistry software research methods.

B. RULE-BASED THEORY AND MASS SPECTROSCOPY KNOWLEDGE SYSTEMS

RBT requires that all candidates belong to a specific class, or family, of organic compounds. Furthermore, the elements of this class in question must be studied very well under an MS profile and their characteristic features and peculiarities known and understood. A computer simulation of their mass spectra can be performed using known fragmentation rules which are typical and characteristic of the particular family of structures.

The user can introduce the fragmentation "receipts" into the SES interactively during the computer session. Peculiar shortcomings of HOT can be eliminated with this approach because when dealing with molecules that are structurally very similar the HOT-predicted spectra are not discriminating enough to allow reliable structural assignments. The user-known MS rules, on the contrary, are very specific for the investigated class of compounds (a series of steroids, for example) and lead to much more detailed spectra, having a higher discriminating power. There are many procedures striving to condense fragmentation rules in MS spectroscopy. This is a kind of reverse structure elucidation: it looks more like structure confirmation, at least in the beginning. What is the strategic meaning of this endeavor? From molecules of known structure and from their spectra we see that better rules for an automated interpretation of unknown spectra of the same class can be established. The major task of a programmer is now to increase the knowledge of the computer system. What ultimately happens is a transfer of knowledge from man to machine and vice versa: an iterative and interactive acquisition and enhancement of knowledge.

Indeed, in these days of ever-expanding computer chemistry research a new branch of software research and application has been conceived: the expert systems. They not only contain general rules that they apply autodeductively, but also a collection of solution schemes, of previously encountered situations, of human-inferred knowledge: the knowledge base. They are called expert systems because they look like robotic experts in their field of action, but they are heavily dependent on the kind of knowledge base given to them by man. In a way they are somewhat more limited in range of action compared to truly pure autodeductive systems, which, relying on few general rules, are capable of extending their exploration to the borders of the problem space. This limitation is certainly counterbalanced by their greater specificity.

It is outside of the scope of this book to enter into a discussion about expert systems; they are still too scattered over a number of specific applications and too different in the methodology of realization to be of pedagogical and informative value in this specialized and narrow computer chemistry context. Still, unclarity reigns over what deserves the name "expert system" and what does not.

VI. TESTING USING SIMULATED CHEMICAL TRANSFORMATIONS

The program REACT[15] embodies another philosophy for the evaluation of candidate structures in the TESTING phase of an SES. It was also developed within the DENDRAL project. Although the latest advances in SES development are forcing a steady reduction in improbable candidates through more refined spectral interpretation tactics, an inference machine analyzing the role of candidates from a completely different perspective is instrumental to the improvement of the self-judging quality of an SES. This intentionally different, but complementary, perspective resides in a "chemical" testing of the candidates.

The candidates surviving previous testing checks are recorded on a file and are submitted one by one to REACT, where they each undergo a simulated chemical reaction. The simulated chemical reaction should (whenever possible) be chosen tactically to reproduce the one (or more) specific laboratory reaction(s) which proved to have the highest discriminating power. By this we mean that a given type of reaction can give rise to a set of products belonging to a larger number of different structural classes than another, less discriminatory type of reaction.

For example, a laboratory chemist, collaborating with a computer chemist, performs an oxidative degradation reaction on an unknown compound (the educt). Suppose that two reaction products are isolated: an aldehyde and a ketone. The same reaction, properly coded, is applied to the available SES-generated candidates. The simulated "reaction products" are analyzed structurally and visualized in the usual way. Only those candidates that have a suitably positioned double bond and yield an aldehyde and a ketone can be saved; all others must be deleted. The definition of the "reaction" occurs in a formal manner using the same metalanguage of CONGEN. If the chemist wishes to define a substructure apt to undergo oxidative scission, like $C=C$, he has to give the following input:

REACTION NAME: oxcleavage   *(user-defined name)*

SITE: *(where the reaction must occur)*
>chain 2   *(link two carbon atoms)*
>join 1 2   *(make a double bond)*
>hrange 0 2 0 2   *(minimum/maximum number of hydrogen atoms)*

>draw atnamed

SUBSTRUCTURE OXCLEAVAGE:

$C=C$

>done

OXCLEAVAGE DEFINED   *(reaction definition terminated; O.K.)*

This formal reaction is applied to all candidates that have at least one reaction "site" available, as defined in the input. However, not all candidates with a $C=C$ substructure lead to the expected products. Only some candidates will have a double bond located such that its cleavage yields exactly one aldehyde and one ketone.

This reasoning is explained with the help of four classes of candidates: A, B, C, and D (Figure 9). If OXCLEAVAGE acts on the topological structures of the educts, other classes of products are obtained: A', B', C', and D'. Class A candidates do not yield two

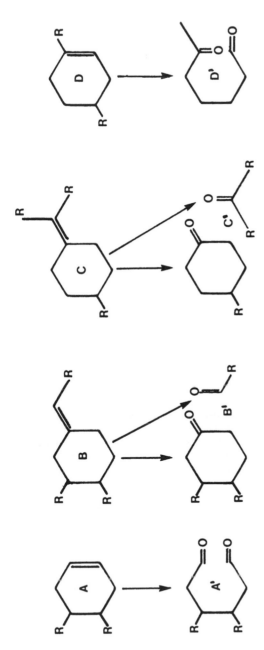

FIGURE 9. Application of a simulated oxidating cleavage reaction to four classes of candidates helps in determining the correct structural class, which must give rise to an aldehyde and a ketone. Only candidates of class B are consistent with this requirement.

products and can be eliminated. Class D candidates indeed yield products with aldehydic and ketonic substructural features (D'), but they also violate the constraint of forming two final products per candidate. Class C educts form couples of products (C'), but they are both ketones. Finally, two products are generated from class B, one aldehyde and one ketone. Only members of class B are candidates that can be retained for further testing (if necessary). All others, in contrast with the experimental result, can be pruned from the file.

This very last theme brings us directly into what may be the most fascinating area of computer chemistry. Chemistry primarily means reactions. Chemical reactions, especially those in the complex world of organic chemistry, are still a challenge for the human mind — its fantasy, its creativeness, its memory. The concluding chapter of this book will present the excellent achievements of researchers who dared to and succeeded in modeling organic chemistry with a computer.

REFERENCES

1. **Smith, D. H., Ed.**, *Computer-Assisted Structure Elucidation*, ACS Symp. Ser., Vol. 54, American Chemical Society, Washington, D.C., 1977.
2. **McLafferty, F. W. and Staufer, D. B.**, Retrieval and interpretative computer programs for mass spectrometry, *J. Chem. Inf. Comput. Sci.*, 25, 245, 1985.
3. **Carhart, R. E., Varkony, T. H., and Smith, D. H.**, Computer assistance for the structural chemist, in *Computer-Assisted Structure Elucidation*, ACS Symp. Ser., Vol. 54, Smith, D. H., Ed., American Chemical Society, Washington, D.C., 1977, 126.
4. **Lindsay, R. K., Buchanan, B. G., Feigenbaum, E. A., and Ledereberg, J.**, *Applications of Artificial Intelligence for Organic Chemistry: The DENDRAL Project*, McGraw-Hill, New York, 1980.
5. **Munk, M. A. and Farkas, M.**, Computer-assisted structure analysis, *Mikrochim. Acta*, 2, 199, 1987.
6. **Woodruff, H. B. and Munk, M. E.**, Computer assisted interpretation of infrared spectra, *Anal. Chim. Acta*, 95, 13, 1977.
7. **Shelley, C. A., Woodruff, H. B., Snelling, C. R., and Munk, M. E.**, Interactive structure elucidation, in *Computer-Assisted Structure Elucidation*, ACS Symp. Ser., Vol. 54, Smith, D. H., Ed., American Chemical Society, Washington, D.C., 1977, 92.
8. **Funatsu, K., Miyabayashi, N., and Sasaki, S.**, Further development of structure generation in the automated structure elucidation system CHEMICS, *J. Chem. Inf. Comput. Sci.*, 2, 18, 1988.
9. **Yamasaki, T., Abe, H., Kudo, Y., and Sasaki, S.**, CHEMICS: a computer program for structure elucidation of organic compounds, in *Computer-Assisted Structure Elucidation*, ACS Symp. Ser., Vol. 54, Smith, D. H., Ed., American Chemical Society, Washington, D.C., 1977, 108.
10. **Funatsu, K., Del Carpio, C. A., and Sasaki, S.**, Automated structure elucidation system — CHEMICS, *Fresenius Z. Anal. Chem.*, 324, 750, 1986.
11. **Funatsu, K., Susuta, Y., and Sasaki, S.**, Application of IR-data analysis based on symbolic logic to the automated structure elucidation, *Comput. Enhanced Spectrosc.*, in press.
12. **Steyn, P. S. and Vleggaar, R.**, A reinvestigation of the structure of monochetin, a metabolite of *monochaetia compta*, *J. Chem. Soc. Perkin Trans. 1*, 11, 1975, 1986.
13. **Werli, F. W. and Wirthlin, T.**, *Interpretation of Carbon-13 NMR Spectra*, Heyden & Son, Philadelphia, 1976, 160.
14. **Gray, N. A., Carhart, R. E., Lavanchy, A., Smith, D. H., Varkony, T., Buchanan, B. G., White, W. C., and Creary, L.**, Computerized mass spectrum prediction and ranking, *Anal. Chem.*, 52, 1095, 1980.
15. **Varkony, T. H., Carhart, R. E., Smith, D. H., and Djerassi, C.**, Computer-assisted simulation of reaction sequences: applications to problems of structure elucidation, *J. Chem. Inf. Comput. Sci.*, 18, 168, 1978.

Chapter 7

COMPUTER SIMULATION OF ORGANIC REACTIONS

I. INTRODUCTION

Since the birth of organic chemistry as a science dealing with the synthesis and char-acterization of molecules containing a carbon skeleton, two fundamental questions have usually plagued every experimental organic chemist:

- How can I synthesize a given target molecule?
- Given a certain substrate molecule, how will it react under given conditions?

The first question calls for a retrosynthetic approach to a synthesis problem, i.e., we have a synthesis design problem; the second question requires a forward search approach, i.e., deals with a reaction prediction problem.

It is evident the these two paramount questions will receive different interpretations depending on whether they are formulated in an academic or in an industrial environment. In a university laboratory, priority is given to a successful synthesis of an interesting, structurally novel, and challenging target molecule, disregarding yield at the beginning; much interest is also devoted to the development of new reaction schemes. The creation act is the chemist's goal; the acknowledgment of the intellectual achievement is his reward.

In industry, which has a profit-oriented organization, the molecule of interest must be obtained at a high yield at the lowest possible cost. Different considerations dominate here, and the "beauty" of a synthetic route experienced in an academic world might easily be lost by a brute high-temperature catalytic conversion in a production plant. Furthermore, the number of compounds synthesized in industry is extremely high: to obtain one phar-macologically active compound which passes all mandatory tests for release on the market, many hundreds or thousands of different species must be synthesized. Sometimes this pro-cedure is not different from a wild, random search for some new lead compound, which might introduce a new class of drugs and then initiate a more systematic investigation.

The questions above can therefore be slightly modified into the following queries:

1. For synthesis design problems:
   A. How can I find a new synthesis (at any cost) for a given molecule?
   B. How can I find the best synthesis (at controlled costs) for a given molecule?
   C. How can I find any synthesis for a new molecule?
2. For reaction prediction problems:
   A. Which reaction products will be generated from certain educts? (no primary interest in product yields)
   B. How can I best obtain a specific product from given educts? (strong interest in the yield of one or more specific products)

These goals can be rephrased to make their most intrinsic strategic consequence more evident. If one wishes to synthesize a molecule over a new synthesis route, whatever the final yield will be, then the novelty lies in the route and not in the molecule. The chemist strives to demonstrate that he has found a new method to assemble chemical subunits to obtain a final structure. Isolation of even 100 mg of final product, just enough for a series of analytical structural elucidation tests (possibly supported by SES sessions), can mean full success and can be proof that the synthesis strategy works. A novel sequence of synthetic steps can itself be analyzed further logically. The novelty can root in the discovery of an unprecedented reaction scheme, a really new (and, hence, previously unknown) pattern of bond breaking and bond making for a given set of atoms involved in the reaction. This original chemical transformation would probably deserve the name of its inventor, and in fact a large number of reactions are known to us as name reactions. A different situation occurs when a novel combination and sequencing of known reactions is established to synthesize a known target molecule. This is a valid intellectual performance as well, as it can be of great interest for its simplicity, its inexpensive materials, its overall speed, and many other considerations of that kind. A similar line of reasoning can be applied to forward search problems. If the investigator formulates on paper the expected products of a certain known reaction scheme R that he applies to substrates which are slightly different from the standards on which the reaction R was derived, he certainly is taking some risks. We all know a certain reaction that "always worked before", which upon change of substrate unexpectedly no longer proceeds or, if it does proceed, delivers products different from those "orthodoxically" expected. After an in-depth investigation of the studied system, the chemist might at this point be able to rationalize the formation of such products and formulate a new reaction model.

On one hand, this can be very disturbing to the organic chemist engaged in a long, multiple-step synthetic approach to a difficult target. Such an unforeseen substrate behavior may heavily undermine his whole synthesis strategy. On the other hand, in luckier events it can provide an insight into a new, exciting class of chemical reactions.

The latter case is normally welcome in an academic environment, whereas the former case can seriously embarrass both the university chemist as well as the industrial chemist, who is tightly bound to the attainment of a specific product. What role does man play in the evolution of all these situations, which are so common in daily laboratory work?

First, so far there is not a complete and general theory on chemical reactivity; second, we cannot count on either an attainable limit to the number of conceivable reaction paths or an absolute number of organic molecules, the known and those still to be unveiled. No human mind can cope with such a wide problem space, such a large number of problem states and molecules, and such intricate and manifold connecting paths. In contrast to structure elucidation, where the interpretation of peak patterns and the assembly of the substructures found must obey strict rules, the feasibility ranges and transformation rules of a "chemical reaction game" are very blurred and flexible. Lacking rigid rules, man has two resources through which he expresses his chemical work and ideas: memory and intuition.

A. THE CONNECTION BETWEEN CHEMICAL MIND AND COMPUTER REASONING

Man's chemical knowledge is first accumulated during his educational period and is strongly influenced by the kind of textbooks he reads, by the major trends in organic chemical research typical of the institute(s) in which he operates, and by the particular way in which his teachers promulgate specific subjects. His knowledge is successively deepened and specialized in his own area of research.

Memory is the mental framework that enables man to formulate reasonable reaction schemes and synthesis strategies. This is so for every individual. Like a computer, we acquire, store, and recall pieces of information from our brains when necessary. The central link between visual perception, memory allocation, and, later, retrieval of data is governed by the recognition of images, geometric patterns, and situations (e.g., one particular event happened in one particular laboratory on one particular day). The "chemical memory" of a chemist works through the stimulus coming from a certain recognized known chemical symbolism, eliciting the issue of correlated quanta of information. For example, seeing the alphanumeric strings

$$\{Br-Br, \quad RHC=CH_2\} \qquad (A)$$

a chemist immediately associates to "bromine" and "alkene" the bromine addition reaction of double bonds. The visual perception of symbolism A is mentally correlated to another symbolism,

$$\{RHBrC-CH_2Br\}$$

which is stored somewhere in the chemist's memory, ready to be activated under the right circumstances.

When the researcher is planning a possible synthesis route for a target compound **T**, his memory activity is to individuate certain substructures inside the molecular frame which are correlatable to his personal chemical knowledge base, i.e., to his memory content. The visual perception of a given substructural entity inside of **T** will in specific cases remind him of a particular reaction scheme, such as a known name reaction, or of a reaction class that fits his purposes.

If memory records are insufficient to satisfy the requests, he might consult more "stable" memories (e.g., books or chemical data banks). It is essential to understand that in this context all walks (reaction steps) over the edges of the graph connecting the target to every possible precursor evolve inside a known subspace of the global problem space. The only distinction between human memory and computer memory is that the former is fallacious and scarcely extendable, the opposite being true for the electronic machine. If we were confronted with the tedious problem of deriving a suitable sequence of different reaction steps to obtain a given **T**, another powerful human capacity could intervene sometimes to provide a sudden "afterburner" effect: it is intuition. A sudden jump in knowledge, not based on memorized facts, but virtually synthesized during the intellectual effort, can be called a successful or enlightening intuition. Intuition brings forth solutions to problems which are beyond the mere rational reach of the investigator, as it seems to "explode" on its own in the human mind. Unfortunately, intuitive progress of knowledge, like the discovery of a new reaction mechanism or the conception of a new, unthought of reaction sequence, is rare.

The advantage of intuitive problem solving is that it is the most powerful way leading out of the known problem subspace into a chemical no-man's-land in which pioneer work still waits to be done. Novel models and different, original concepts can thus be formulated for the first time. The mental attitude of an organic synthetic chemist in some aspects parallels the creative work of an artist, especially of an architect. He has a spectrum of known style elements which he combines into an original architectonic realization. The personal touch, the individual fashion, can give rise to a valuable artistic contribution, a mixture of classic knowledge and personal invention. The intuitive breakthrough in architecture, on the other hand, means the creation of a new style period in which completely new formal architectonic elements and their descriptive language are established.

The organic chemist forges in his mind 2-D and sometimes 3-D "architectonic" microentities, the molecules, which descend from abstract formal elements — the known substructures, the reaction schemes. At the very moment of an intuitive realization of an unprecedented reaction scheme, i.e., of a new formal element of the "chemical language", he very much resembles an artist. All such complicated relationships between man, knowledge, memory, intuition, formal chemical elements, and rules have been projected into a computer to create AI modules which are flexibly named computer-aided synthesis programs (CASP).[1-3]

CASP probably represent the most daring attempt to develop computer programs that "think" in chemical categories, especially because (as mentioned above) there are no strict and orthodox rules for chemical reactivity. Two main philosophies of conceiving a synthesis design system (SDS) exist. First, we have the SDS based on reaction libraries, thus emulating human memory and attempting to model chemical reasoning according to a predefined knowledge base; leading examples of this philosophy are programs like SECS,[4-6] LHASA,[7,8] SCANSYNTH,[9] and SYNCHEM-2,[10,11] just to cite the principal and most classic achievements in this area. Second, there is the SDS based on formal generalized reaction generators, devoid of any reaction data base. These systems tend to model intuition and are potentially capable of entering unexplored regions of the problem space.

II. SYNTHESIS DESIGN SYSTEMS BASED ON REACTION LIBRARIES

It must be clear that an SDS exploiting available information about organic chemical reactions can explore only the problem space inside the boundaries spanned by the available knowledge base. It contains thousands of known chemical reactions which are properly coded as formal transformations. No new reaction scheme will ever be found; instead, novel reaction sequences and alternative synthesis strategies can be generated. To clarify this assertion the Diels-Alder reaction is used as an example. If this particular reaction is "forgotten" by the programmer of an SDS, meaning that it is not coded explicitly within the program, the computer will never find it.

This kind of SDS was the first to be programmed; this occurred in the prototype OCSS,[12] as it is more proximate to the thinking patterns of a chemist, who relies heavily on his memory. Obviously, the enormous advantage of a computer is that it never "forgets", not even one of the many thousands of reactions located inside its magnetic memory.

The methods and principles that researchers used to design different prototypes of such sophisticated program systems shall be presented according to their invariant features, which are common to the class of synthesis design systems utilizing reaction libraries.

A. STRUCTURE AND TERMINOLOGY OF A SYNTHESIS DESIGN SYSTEM

The synthesis tree in Figure 1 schematically expresses the antithetic search for synthesis routes leading from certain sets of starting molecules S_i to the final target molecule T (the goal). Except T, which is the input molecule, all nodal problem states P_{ij} and S_i inside of the problem space portion embraced by the synthesis tree are predicted molecules; the chemical reactions R_{ik}, the transformations, are the connecting edges. The edges are abstract representations of rules of chemical transformation of a node into another node. Formally taking R as an operator we have

$$R^0(T) \rightarrow P^1, \quad R^1(P^1) \rightarrow P^2, \quad ..., \quad R^n(P^n) \rightarrow S \tag{1}$$

The retrosynthetic search implies that a computer "thinks" backward, i.e., from T back to the set of starting materials S, applying transformations which are in reality retrotransformations because the real synthesis in the laboratory will proceed from S to T. The nodes P are also called subgoals or precursors. The number of levels of the tree establishes the depth of the search.

The strategy behind a synthesis tree is that T is dissected into simpler, constitutionally smaller precursors. The central problem is the selection and application of suitable transformations R. The general process of generation of a synthesis tree contains (among many others) the following important steps:

1. Data input — the structure of T is input either graphically or alphanumerically by the user.

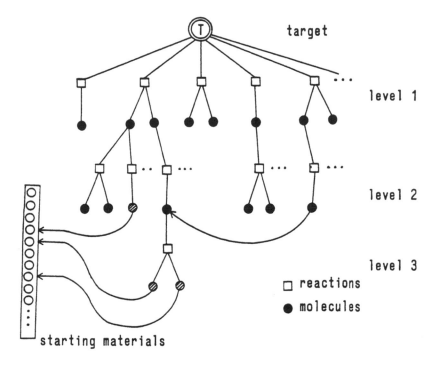

FIGURE 1.   The synthesis tree.

2. Cognition — the structure is analyzed, or better yet perceived, by the computer in all its chemical features, such as heteroatoms, rings, aromaticity, functional groups, strategic bonds, etc.
3. Strategy — the system autodeductively individuates strategies for retrosynthetic dissection of **T** (and later, of the various **P**s).
4. Execution — the transformations are applied and the nodes generated.
5. Check — if **P** = **S**, or if a user-given depth boundary is reached, the simulation process stops.
6. Structure output — the generated structures are visualized and plotted for the user's interpretation.

Points 1 and 2 have been treated in the section about representation of molecular structures.

The perception of specific substructural features or functional groups, from the simplest (like a carbon-heteroatom bond or a double bond) to very large ones, is the driving force in this particular concept of an SDS. Because the computer has been instructed to hatch its suggestions consulting a reaction library and because the feasibility of a certain retroreaction may depend on the simultaneous presence of more than one of these elementary functional groups, a solid, thoroughly dedicated search is necessary.

This specific search tackles the problem of individuating some structural patterns, joint or disjoint, inside of **T** (and, of course, of any **P** in later levels) which must be matched to some reaction scheme contained in the reaction library. If a correspondence is established, the specific library reaction can be rendered operational.

Thus, if in the cyclohexene structure shown here below, for example, the SDS localized just the isolated double bond as a functional group, the first level of precursors would probably be only an expression of several methods of forming a double bond (in the synthetic direction), as shown in scheme A.

Conversely, if the computer comprehended the full essence of the six-membered unsaturated cyclic system, it could recall from the reaction library the transformation equivalent to a retro-Diels-Alder reaction (scheme B above). This retrosynthetic step is significantly superior in its strategic quality to the various structural modifications coming from a localized action on a double bond: in A the global compexity of the precursors is similar to **T**, but in B the simulation problem is already solved on one level with the educts S_1 and S_2, two very simple molecules which do not require further retrosynthetic analysis.

We see that, in projecting a library-oriented SDS, importance must be given to modules responsible for the effective perception of structural features which allow sound planning of bond dissection strategies and tactics. This can happen upon direct recognition of functional groups or after intervention of functional group interchange routines. The bonds following these requirements are called strategic bonds.

B. TRANSFORMATIONS (R)

The functional group interchange[13] (FGI) is an element of one of the three principal classes of transformations allocated in separate libraries of the SDS. The FGI is conveniently used to modify a certain functional group such that the successive step leads to a fundamental decomposition of **T** (or of any **P**); an FGI therefore generates explicit strategic bonds from their latent forms.

The three main classes of retroreactions implemented in the major SDS as transformations **R** are the following:

1. Class 1 — to this class belong all **R**s requiring a pair of functional groups connected by a path in the molecular graph. Three subcalsses can be defined (see Figure 2).
    A. **R** disconnects the path between the two functional groups.
    B. **R** forms a new path between two functional groups.
    C. **R** modifies the functional groups without altering the path.
2. Class 2 — the transformations here involve a single functional group, yielding the following subdivisions:
    A. **R** breaks the path leading to a functional group.
    B. **R** generates a ring.
    C. **R** is of FGI type.
3. Class 3 — this class contains **R**s which depend on the size and the functional substitution pattern of rings (see Figure 3). We have divided them more explicitly into the following subclasses:
    A. **R** dissects a ring.
    B. **R** forms a ring.
    C. **R** causes a rearrangement.

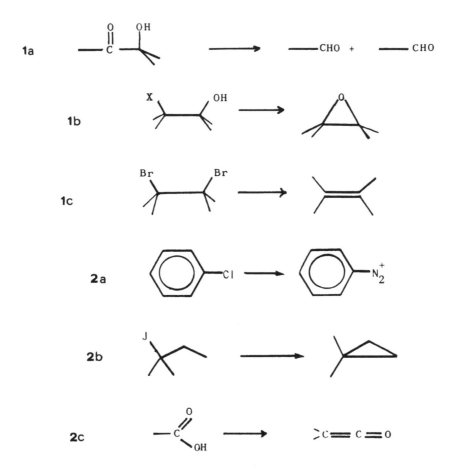

FIGURE 2. Examples of class 1 and class 2 transformations: the former require the presence of two functional groups connected by a path; the latter require only one functional group.

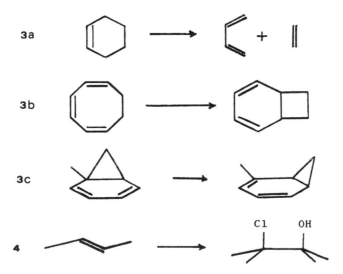

FIGURE 3. Examples of class 3 transformations, specifically dependent on ring size and on its substitution pattern, and class 4 transformations, which generate two functional groups.

Another important class of tranformations must be mentioned, the one that attaches functional groups to the preexisting skeleton (class 4).

There are many other classes of **R** that will not be discussed here. Each one represents a certain family of chemical transformations. The members of this family are the real organic reactions, with their typical names, that at one time we have all studied and become familiar with.

The transformations of class 3 are quite important because the antithetic path connected to a ring system has a high strategic value. The formation of a ring in the synthetic direction can have a decisive consequence for the entire synthesis, especially if **T** contains more than one ring. Eventually one should consider the difficulty that a chemist encounters in deciding the order in which the different rings should be synthesized. This difficulty is presented in a reverse manner to a retrosynthetic program, which must evaluate which ring would best be dissected first to allow an optimal synthesis at a later time. The computer thinks ''backward'' here. The number of possibilities for applying the described classes of transformations to a substrate is extremely high , and many are not convenient. Many FGI transformations, for example, will alter the functional pattern of **T** without offering any valuable contribution to the solution. In addition, some transformations might be applicable at different sites on one and the same molecule, but with completely different effects on the quality of the retrosynthetic simulation.

For these and analogous reasons, each SDS incorporates AI modules which aim at establishing evaluation criteria for the goodness of the autodeductively derived reaction paths.

C. EVALUATION STRATEGIES AND TACTICS
1. Strategic Bonds

A common method to obtain useful retrosynthetic proposals from a computer involves the role of strategic bonds. These are generally topologically important bonds which, if broken by a certain **R**, result in a considerable change in the topological structure of the processed molecule. The change will be reflected dramatically on the internal representation of the molecular structure. Consider, for example, the retrosynthetic study of compound **A**, containing a given heteroatom X in an endocyclic as well as in an exocyclic position. Let an appropriate **R** act on **A** such that X is removed from **T**.

Two results with different strategic weights are obtained. The left precursor shows an intact basic structural core of **T** due to a limited, exocyclic action of **R** in the mode of the class 2a FGI transformations. The precursor depicted on the right is generated from a transformation which modifies the structural aspect of **T** profoundly: two cycles are eliminated, the one containing X and the virtual decaatomic cycle. In both cases, one and the same hypothetical **R** was acting on an −X− substructure. Thus, the two bonds departing from the endocyclic X certainly are to be considered strategic.

Also, feasibility and cost criteria for a synthesis have been considered in the identification of strategic bonds. With the disconnection of strategic bonds splitting **T** into approximately equal halves, it was shown that in this case the total synthesis of the target required a minimal number of levels and of precursors.[14] Another accepted technique is the modification of functional groups through FGI operators, groups that in their present forms do not allow any immediate strategic transformation. When the computer searches the target for substructures that match the requisites of a certain class 1 transformation **R** (two functional groups necessary), it may happen that the first functional group is found to be valid, but the second results in a mismatch, thus preventing the activation of **R**. At this point an FGI transformation can be applied on the blocking substructure to convert it into the missing but requested functional group. The bifunctional **R** can now become operative. We can follow this technique on the example below, in which a simple conversion of a hydroxylic unit into an oxo unit paves the way to an intervention by a strategic, bifunctional **R**, which cleaves **T** into two simple starting materials.

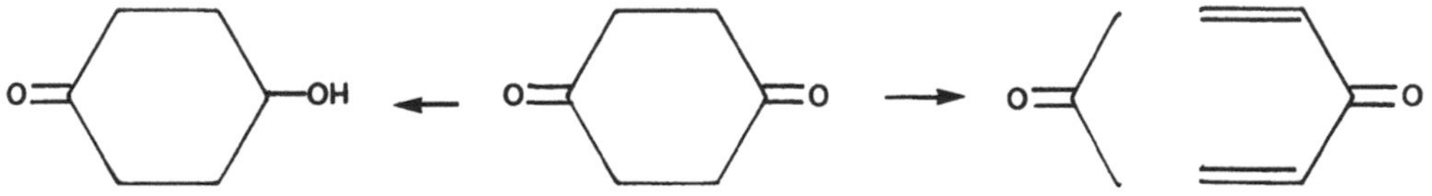

Several sources of heuristic information help the SDS in localizing strategic bonds or strategic groups in order to generate the simplest synthesis tree possible that still has chemical significance. Cost functions are computed for each transformation depending on whether the structural environment around the site on which a given **R** is meant to operate is more or less favorable. The global cost of any transformation **R** is given by taking into account the presence or absence of other functional groups or substructural elements favoring (positive score) or disadvantaging (negative score) the action of **R**. For example, the chemical equation **a** shown below represents the transformation of the aldol condensation, to which a standard score is attributed. If (like in equation **b**) a C=O group augments the acidity of the α-CH$_2$ hydrogen atoms, a positive score is added to the standard (bonus). On the contrary (equation **c**), the presence of a discordant functional group like $-$NH$_2$ strongly inactivates the enolization of the nucleophilic methylene carbon atom, and a negative score (malus) is added.

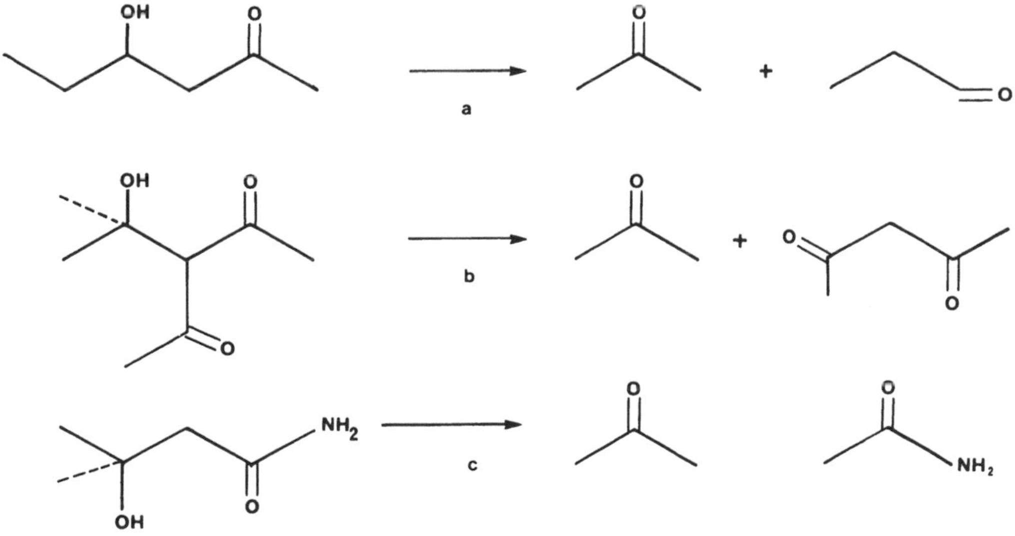

Before correlating a certain **R** to the available transformable substructural sites, a ranking of all possible transformations is set up with the calculated scores to discern the best one, to which preference is given. Retroreactions with a score lower than the permissible threshold value are discarded.

2. Recognition of Functional Groups

SDSs designed on reaction libraries have collections of thousands of coded organic reactions. Each one of these reactions can be activated as soon as the system finds in **T** the proper structural patterns corresponding to the matching pattern of the one specific transformation. The particular R encoding, say, the Beckmann rearrangement will be addressed after the identification of the $-NH-CO-$ substructure.

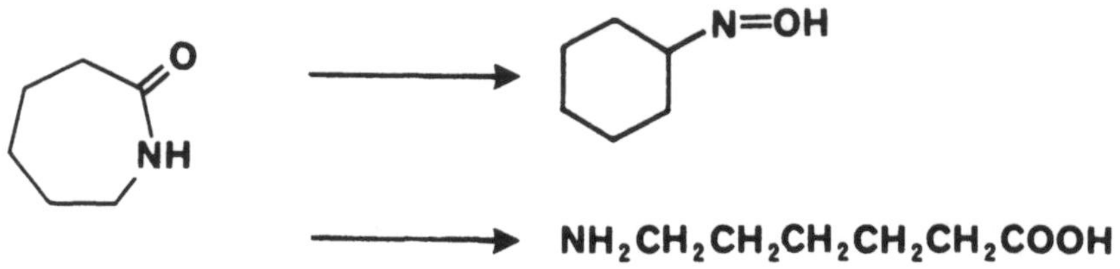

Also, an **R** for a simple hydrolysis generating an amino acid is considered. The matching process implies two complementary steps: the choice of a methodology for the human-established codification of a real chemical reaction into a computer-readable form, and a formalism for its internal representation. The internal representation is substantially a formal description of substructures.

We have learned that any molecular structure in its topological form can be decomposed into an arbitrary number of functions. From them a functone connectivity matrix (FCM) can be constructed. For example, let us take structure **A** to represent any **T**.

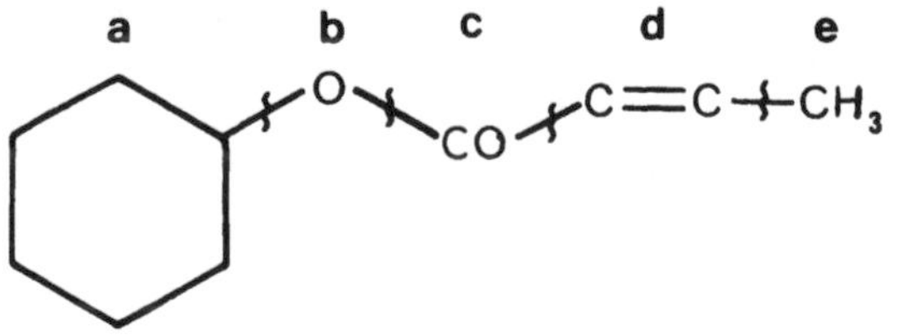

Structure A contains the functones **a, b, c, d,** and **e**. FCM(A) has the following form:

	a	b	c	d	e
a	0	1	0	0	0
b		0	1	0	0
c			0	1	0
d				0	1
e					0

FCM(A) =

There exists an infinite class of structures of type **B**:

$$\$1-O-CO-\$2 \qquad \textbf{B}$$

where $\$1$ and $\$2$ are symbols for generic substituents. The corresponding FCM(B) is

	$\$1$	b	c	$\$2$
$\$1$	0	1	0	0
b		0	1	0
c			0	1
$\$2$				0

FCM(B) =

B represents a synthon. A synthon is a small synthesizable unit, coded by its specific FCM(synthon) and stored in the synthon file of the reaction library. The matching algorithm seeks out coincidences between the current FCM(synthon) and the FCM(A). There can be more than one FCM(synthon) that correctly matches submatrices of an FCM(target). However, this does not automatically guarantee that the related reaction (transformation) is unproblematical in its feasibility. In fact, the searching algorithm also would localize the $1–CO–CH=CH–$2 synthon, which formally would call for a retro-Michael addition in **T**, yielding Ph–O–CO–CH$_3$ and O=CH–CH$_3$. This reaction, however, is pruned by a further check: in synthon B the oxygen atom linked to $1 is a neighbor of the same carbonyl group of synthon $1–CO–CH=CH–$2. The two synthons clearly overlap, and C=O is the common substructure. The additional oxygen forms an ester group, which partially inhibits the acidity of the methyl hydrogens, necessary to generate the attacking nucleophile. In the presented situation, the Michael-addition transformation would receive a negative score (scaled to a standard set of reagents). A better approach to **T** is embodied by another kind of retroreaction in which the ester group is no longer a negative interfering entity. The presence of deactivating interfering substructures is established by a control algorithm using a bit string technique. Every position of a bit inside a special control word corresponding to a certain reaction is a logical label for a certain functone and acts as its representative. Imagine, by the following example, that the role of a generic halogen atom X is attributed to the first bit of an n-bit word, the role of a carbonyl group to the second bit, the one of an alkyl group to the third, and so on. The bit string describing a given reaction is constituted by "1" and "0" bits, indicating that a particular functone either favors or handicaps the reaction, respectively. If, for example, the control word belonging to the retroesterification of a generic synthon **B** was composed of the bit string

Functone	X	CO	Alkyl	Aromatic	C=C ...
Bit	0	1	1	1	1

the control algorithm would know that a halogen in position $2 of synthon **B** would disfavor the cleavage of the strategic bond O–C (the bit related to X is FALSE) because a reaction like

$$O–CO–Cl + OH^- \rightarrow O–CO–OH + Cl^-$$

would be concurrent. The control algorithm receives a bit string describing the real structure of **T**, molecule **A** in our example. The control bit string for the submatrix of functones around the substructure –O–CO– in **A** (which has been positively matched to the synthon of class **B**) is

Functone	X	CO	Alkyl	Aromatic	C=C ...
Bit	0	0	1	0	1

This logical representation simply says that the partners of the substructure –O–CO– in **A** are an alkyl group (the cyclohexane ring) and an alkene unit (the propenyl residual). The comparison of the bit string of the submatrix and the general bit string for synthon class **B** shows that the substructures neighboring the synthon-equivalent substructure in A are not in contrast to those required for the acceptance of the retroreaction linked to the synthon. If, on the contrary, the bit string of the synthon in A described, for example, an ester derivative of dichlorocarbonic acid (alkyl–O–CO–Cl), its bit pattern would be

Functone	X	CO	Alkyl	Aromatic	C=C ...
Bit	1	0	1	0	0

The presence of a "1" in the position of a generic halogen, if compared to the equivalent position of the synthon **B** string, reveals a discrepancy: a *false* entry (synthon **B**) and a *true* entry (in the example substructure) give *false* in the logical **AND** operation, with a possible rejection of the reaction.

To summarize, this bit string technique compares the requirements of general retro-reactions (expressed by the bit strings of classes of general synthons) with the actual environment of the corresponding synthon-equivalent substructure of **T**. The consequence of the matching mechanism is an empirical method used to evaluate the above-mentioned cost functions (i.e., scaled probability).

On similar bit strings the computer performs the perception and manipulation of all those structural features which legitimate a computer-driven simulation of chemical events. These features are called qualifiers. They are concatenated ingeniously in the program prescriptions of the chemical transformations. Qualifiers create a detailed identikit of the molecule, emphasizing certain characteristics of atoms, bonds, or larger groups. Qualifiers are descriptors invented by the chemist which always correspond to our traditional, accepted semantic description of a molecule. The following qualifiers are self-explanatory: HALIDE, DBOND, HETERO, QUATERNARY, EXOBOND, ALFA-TO, and so on. It is always the computer chemist who decides about the introduction of the necessary number of qualifiers in order to assure a reliable, more flexible, and realistic reproduction of chemical reality. We want to show how the manipulation of logical qualifiers can extract important information from a molecular structure, information that catalyzes the application of a certain transformation R. As usual, let us construct a short example.

The cyclic structure, whose atoms are arbitrarily numbered

must be evaluated by the SDS in order to locate within the representation of an aldol retroreaction the atom(s) capable of donating protons. Let us invent the following qualifiers and generate the full logical description

		atom number					
	1	2	3	4	5	6	7
OXO =	(1	1	0	0	0	0	0)
HETERO =	(1	0	0	0	0	1	0)
CDB =	(0	1	1	1	0	0	1)

CDB labels carbon atoms with a double bond; in addition, n different bit vectors (n = the number of nonhydrogen atoms) are defined (ATTACH) that label the atoms bonded to a central atom. For example, ATTACH for atom 2 would be (1 0 1 0 0 0 1).

A possible informal scheme of instructions (comments in italics) could be

ALDOL:
  DO FOR ALL ATOMS
  IF OXO(i) AND NOT HETERO(i) = TRUE   *yields (0100000)*
  THEN PROCEED
Only atom 2 is TRUE
  DO FOR i = 2
  IF ATTACH(2) AND CDB(ATTACH(2)) = TRUE   *yields (00010001)*
  THEN PROCEED
Only atoms 3 and 7 are TRUE
  DO FOR j=3, j=7
  IF ATTACH(j) AND NOT OXO(ATTACH(j)) AND CDB(ATTACH(j))=TRUE
  THEN CALL ALDOLTRANSFORM

The last logical equation can be expanded for clarity. For atom 7 we can write

(0100010) **AND NOT** (1100000) = (0100010) **AND** (0011111) = (0000010)

and then

(0000010) **AND** (0111001) = (0000000) = FALSE

For atom 3 we have

(0101000) **AND NOT** (1100000) **AND** (01110011) = (0001000) = TRUE

This last result recognized atom 3 as a potentially valid carbon atom having acidic hydrogens, enabling the call for a retro-aldol transformation.

In all programs using reaction libraries, specially developed metalanguages are used to construct the simulated chemical retroreactions, the transforms. SECS uses the ALCHEM language (**A**ssociative **L**anguage for **Chem**istry), while LHASA uses CHMTRN (**Chem**istry **Tr**anslator). An ALCHEM or CHMTRN sequence of instructions, a transform, is translated by an interpreter program (which is a compiler in the strict sense) into a machine-independent binary description. Every command of the metalanguage can be read by the main SDS program, which normally is written in a higher programming language. The transform is converted into a chain of elementary instructions for the computer. A complete discussion of a chemical metalanguage is certainly outside the scope of this book, but one example of an ALCHEM transform is added below. The English-like metalanguage makes the meaning of the individual instructions easily understandable.

TYPE PAIR
;
;
;;;;;   MICHAEL ADDITION
;
;   ADDITION OF ACTIVE METHYLENE COMPOUNDS TO VINYL-W GROUPS
;
;   W–C–C–CH–W   W–CH   +   C=C–W
;   DONOR   ACCEPTOR
;
;
;
;

```
;REF:    HOUSE, MODERN SYNTH. REACTIONS 595 (1972)
;        BERSMANN, ORGANIC REACTIONS 10, 179 (1959)
;SUB:    (ACCEPTOR) NO2 > SO3R >CN>COOR>CHO>COR AND
;          SO2+X, C=N+R2, SO2N, SO2R, PO(OP)2, CONR2, C=NR, ARYL
;REA:    (DONOR) SAME GROUPS AS ACCEPTOR BUT SEQUENCE OF REACTIV-
;          ITY UNKNOWN CATALYTIC AMOUNTS OF WEAK BASES OR NOTH-
;          ING, ROOM TEMPERATURE (HIGH TEMP FAVORS FOLLOWING CON-
;          DENSATIONS, REARRANGEMENTS AND RETRO-MICHAEL)
;
MICHAELADD
WGROUP WGROUP PATH 5 PRIORITY 45
CHARACTER CLEAVES CHAIN
FGI OK
;   ATOM 4 MUST BEAR ONE ACTIVE HYDROGEN
        IF ATOM 4 IS QUATERNARY THEN KILL
;   W-C-C(=)-C-W NOT SINCE W-C=C= UNSTABLE
        IF DOUBLE BOND IS ALPHA TO ATOM 3 OFFPATH THEN KILL
;   EXOPATH DOUBLE BOND AT ATOM 2 ONLY ALLOWED
;   IF CORRESPONDING BETA-ATOMS BEAR ANY HYDROGEN
        IF DOUBLE BOND IS ALPHA TO ATOM 2 OFFPATH THEN
        BEGIN IF ATOM BETA TO ATOM 2 IS ATTACHED TO ANY HYDROGEN
&       THEN CONTINUE
          ELSE KILL
        DONE
;   CREATE SET (1) CONTAINING PATH ATOMS 2, 3, 4
        IF ATOMS 2—4 ARE ALL CARBON (1) THEN CONTINUE
        ELSE KILL
;   TO EXCLUDE CYCLICATION TO PYRIDINES, ETC.
        IF (1) IS AROMATIC ATOM THEN KILL
;   AR-C(TRIP)C-[COOR,COR] FORM CYCLIC COMPLEX COMPOUNDS
        IF BOND 2 IS DOUBLE BOND THEN
        BEGIN IF ATOM ALPHA TO ATOM 3 IS AROMATIC THEN
          BEGIN IF GROUP 1 IS NOT KETONE THEN CONTINUE
            ELSE BEGIN IF GROUP 1 IS NOT ESTER THEN CONTINUE
        DONE
        DONE
;   OHC-C-C-C-CHO ALDEHYDE IN DONATOR AND ACCEPTOR
        IF ATOM ALPHA TO ATOM 2 IS ALDEHYDE THEN
          BEGIN IF ATOM ALPHA TO ATOM 4 IS ALDEHYDE THEN KILL
        DONE
        ELSE CONTINUE
;   BOND 2 ONRING LEADS TO UNSTABLE PRECURSOR
;   IF GROUP 1 IS KETONE ACYCLOHEXENES UNDERGO SELF-
;   CONDENSATION
        IF GROUP 1 IS KETONE THEN
        BEGIN IF BOND 2 IS ONRING THEN KILL
        DONE
        DONE
;   PRIORITY MANIPULATIONS
        IF WGROUP IS ALPHA TO ATOM 2 (3)
        IF WGROUP IS ALPHA TO ATOM 4 (4)
```

```
;   ADJUSTMENTS FOR DONATOR
        IF (4) IS NITRO THEN ADD 25
        IF (4) IS NITRILE THEN ADD 15
        SET VALUE 1 TO COUNT (4)
        IF VALUE 1 IS EQUAL TO 2 THEN ADD 10
        IF VALUE 1 IS EQUAL TO 3 THEN ADD 10
;   ARYLIC SUBSTITUTION FAVORS REACTION
        IF ATOM ALPHA TO ATOM 4 IS AROMATIC THEN ADD 20
;   ACCEPTOR ADJUSTMENTS
        IF (3) IS ALDEHYDE THEN
        BEGIN MULT BY 10
          DIV BY 7
        DONE
        IF (3) IS KETONE THEN
        BEGIN MULT BY 10
          DIV BY 8
        DONE
        IF (3) IS NITRILE THEN
        BEGIN MULT BY 10
          DIV BY 9
        DONE
;   AMIDE LEADS TO UNDESIRABLE BY-PRODUCTS
        IF (3) IS AMIDE THEN DIV BY 2
;   AROMATIC SUBSTITUTION FAVORS REACTION
        IF ATOM ALPHA TO ATOM 1 OFFPATH IS AROMATIC THEN
        BEGIN MULT BY 10
          DIV BY 7
        DONE
        CONDITIONS SLIGHTLY BASIC
;
        CLEAVE BOND 3
        MAKE BOND 2
END
```

3. Strategic Routes

Strategies for bond selection are a necessary, but not a sufficient, device for extracting the best retrosynthetic pathways from all of those that are formally possible. A strategy for the overall judgment of a full strand of disconnective steps (i.e., the whole chain of connected precursors) must be added. Stated differently, this means that the "quality" of a single branch of a synthesis tree must be quantified. Judging some of these branches to be highly improbable due to particularly inefficient transformations (low yields, difficult reaction conditions) introduces the pruning methodology of selection. Pruning can arise from very differently positioned considerations. The most drastic pruning is reflected in the strategy of limiting the retrosynthetic search to a predetermined number of levels. If the user imposes a backward search over two levels, two generations of precursors are created. The quality of the proposed pathways can be judged reasonably only if enough pathways are generated: this calls for a breadth-first search generating the synthesis tree.

The early pruning method involves the evaluation of the very first step ($T \rightarrow P_{1i}$); reaction paths with a low probability are not allowed to proceed any further. The walk on the tree branch stops here. A large amount of otherwise wasted computer time is thereby saved. The judgment about "quality" contains, of course, a lot of heuristics, but this is quite natural in chemistry. Pruning also can be decided upon by considering the shape of the growing tree. For example, branches of a given shape can be favored. The reason for this approach is that a convergent synthesis is often preferable to a linear chain of consecutive steps. The following three synthesis proposals for **T** contain all seven precursors. They differ in branching complexity and depth, but trees 1 and 2 have an equal number of steps. Supposing a standard yield of 90% for each step, one gets

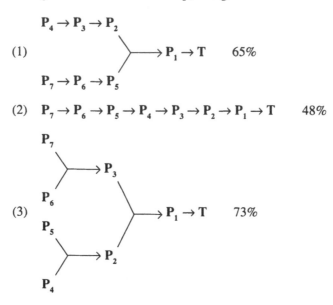

Path 2 is the one with the lowest yield, while path 3 is the one with the shortest depth bound.

Another way to control the generation of a synthesis tree is to force its branches in a specific direction, which in a stepwise manner approaches particular precursors already contained in a large library of readily available compounds. In the SDS SYNCHEM-2, the file of available compounds has more than 5000 entries extracted from the standard catalogs of several chemical supply companies. A compromise between this catalog-oriented pruning philosophy and one free from any such constraints is found in LHASA.[15] A heuristic strategy searches for retrosynthetic schemes that transform **T** into **T'**. This modified target structure can be synthesized via a halolactone educt. The class of all **T'**s contains the so-called strategic intermediates which lead to the respective halolactones through retrosynthetic steps that are known and constant. The difference between the SYNCHEM-2 method and this particular LHASA approach is given in the attempt to obtain known educts (SYNCHEM-2) and known intermediates involving subsequent known reactions (LHASA). The unknown part of the synthesis tree is limited to the paths and nodes departing from **T** to the convergence node **T'**. From **T'** to the various starting materials **S**, which do not have to be identical to commercially available compounds (but could be if such a compound's catalog is connected to the SDS), the pathways are almost rigidly predetermined in their algorithmic expression. Figure 4 shows the convergence of a synthesis tree toward a generic intermediate whose structural patterns are easily converted into a starting halolactone.

The substituents in **T** are checked one by one to identify those which score highest according to the cost functions. Through FGIs and other transformations, the chosen groups (A, F, C) are converted in order to bring **T** structurally nearest to a member of the class of generic intermediates **T'**. The residual substituents (D', E') do not interfere, or they may even enhance the probability of the total conversion.

Various kinds of problems are encountered in the realization of this empirical model.

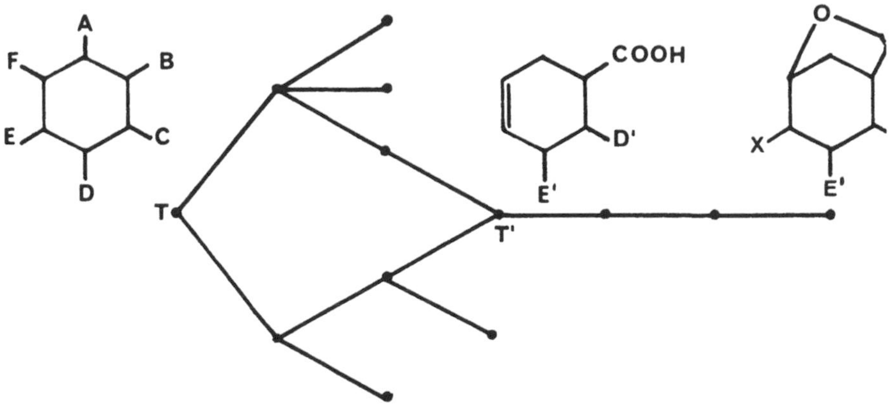

FIGURE 4. The characteristic "converging" tree illustrating the approach based on a strategic halolactone intermediate.

The ring size is determinant, as the halolactonization reaction is easily possible only for five-, six-, or seven-membered rings. The multiple combinatoric possibilities for defining atoms of the substrate as potential points of conversion into halolactonic structural elements (localized matching units) and other hurdles like steric hindrance cause further difficulties for the system. In the course of the simulation process it is not always easy to evaluate *a priori* the steric characteristics of the generated precursors.

Not far from this concept, but supported by a more solid formalism (see below), we find the approach based on chemical distance. It plays an important role in the SDS using formal reaction matrices, which will be discussed later. It can be shown that a matrix representation for a system of molecules before and after the reaction is possible. Let matrix **B** describe the molecular system of educts and matrix **A** the system of products. The difference between the two matrices is the reaction matrix **R**, which is the chemical distance. Since the chemical distance is measured by a matrix **R**, it is possible to compute the global chemical distance for any educt-product couple and, therefore, evaluate each branch of the synthesis tree from the total chemical distance between **T** and **S**. In cases where the structures of the starting materials are known, the computer steers the creation of nodes in the direction in which the chemical distance from given reference molecules is kept lowest.

This is an attractive idea, but computing time requirements when dealing with large molecules make it difficult to realize in practice. A similar but "faster" approach has been geared in the SDS SYNGEN,[16] in which a special algorithm tries to simplify the problem space. Here a "distance function" is used to obtain the minimum number of steps needed to transform a given substrate into a given product. The distance function DF is the net difference between educt and product in the absolute number of hydrogens (Δh) and of heteroatoms (Δz) on each carbon atom i. DF is therefore a measure of the number of virtual steps for a given transformation:

$$DF = \frac{1}{2} \sum_i (|\Delta h_i| + |\Delta z_i|) \qquad (2)$$

Also, LHASA has recently implemented an analog strategic judgment criterion.

Another consideration is of crucial importance: the difference between interactive and noninteractive programs. In interactive programs the user takes a direct part in the simulation process. The pruning is then the result of both precodified rules, known to the computer through the specific man-programmed instructions, and decisions taken in real time by the chemist who sits in front of the terminal. To simplify the man-machine interaction, the SDS

PASCOP[17,18] even offers its own language of communication[19] whose syntax rules are simple enough to be learned rapidly by any chemist. LHASA and SECS, if run in an interactive mode, require the constant alertness of the user during the session.

In noninteractive programs like SYNCHEM-2, internal strategies result from completely precodified rules, and the user has no means of intervening in the generation of the synthesis tree. SYNCHEM-2 has one additional interesting feature that provides for increasing the quality of the chosen strategy: it relies on a constantly expanding knowledge base of chemical reactions. The reaction library is not a list of programs, but a data structure. Intelligence traffic is shuttled between user and SDS through the facility provided by the Knowledge Interchange System. Chemical performance is enhanced by revision of the knowledge base, of the empirical, heuristic, and physicochemical parameters that form the data structure of the reaction transforms. When poor chemistry is modeled by the system, user-suggested additions and improvements are made permanent in the program. This methodology is analogous to the training sequence that characterizes an expert system.[20]

The stretch that we have covered into the recesses of the problem space has brought us to the borders of the domain of human-known chemistry. The various implementations that have been cited have one thing in common: they all constitute a powerful management of known chemistry within an AI perspective. The advantages offered by these SDSs to industrial chemists facing ever-increasing loads of new compounds, patents, and requests for synthetic efficiency and optimization are evident.

A completely different approach, to be discussed hereafter, elongates our imaginary trip to cast a glance on ''the dark side of the problem space''.

III. SYNTHESIS DESIGN SYSTEMS BASED ON FORMAL REACTIONS

A. MATRIX REPRESENTATION OF ORGANIC REACTIONS

Synthesis design systems using reaction libraries only partially exploit the potentially available walks through the problem space around a target. Only what is already known can be used in the simulation. Being tailored for retrosynthetic searches, the prediction of reactions in a forward direction is not feasible with this kind of computer program. Daily experience shows us that new reaction schemes are indeed found in organic chemical laboratories. Somewhere there must be a deep well out of which the chemist extracts new fragments of knowledge about organic chemistry, pieces of a giant puzzle jealously concealed and seldom released to our intelligence. The absence of our private knowledge about a reaction does not mean that this particular reaction does not occur. A Claisen rearrangement is a Claisen rearrangement, independent of our own mental awareness. But how can we program reactions that we do not know yet? This is a logical objection, but we must look at the problem from a different angle. What must be looked for is not so much a huge collection of reactions, but a restricted number of rules which make ''playing'' with atoms and bonds feasible in all possible directions.

The term ''all possible directions'' sounds somewhat crazy because we certainly do not want to do ''all possible chemistry'' in an erratic and random fashion. The emphasis lies on the word ''possible'' as an expression of power, of opportunity. To realize a computer program that has the power of discovering the new a model of the constitutional chemistry is necessary. This model rephrases our chemical knowledge in terms of mathematical entities. The bulk of memorized chemical notions is reduced into a restricted basic set of fundamental formal mechanisms from which all individual differentiations can be inferred.

The usefulness of relevant empirical data will certainly not be neglected, as these data will contribute to setting the width of the search paths when traversing the problem space. It is good to know that ''all possible'' paths are available, but only a limited number of them are needed for a given problem.

1. Ensembles of Molecules and *BE* Matrices

To introduce a mathematical model of constitutional chemistry[21] the following definition is required:

An ensemble of molecules (EM) consists of molecules which can be identical or different. Like molecules, an EM has its own ensemble formula, which is the sum of the molecular formulas of the single species inside of the EM.

The concept of isomerism thus can be generalized immediately: ensembles of molecules are isomeric if they have the same ensemble formulas. From a given collection (A) of atoms of different kinds (*n* carbon atoms, *m* hydrogen atoms, *k* oxygen atoms, etc.) there is a finite number of formally possible isomeric EM(A). A family of isomeric ensembles of molecules (FIEM) is the set of all EM(A). A chemical reaction is the conversion of one EM into another (isomeric) EM. Thus, an FIEM(A) contains the complete chemistry of an EM belonging to a set of atoms A. It should be clear that an EM is not limited to the traditional isomerism concept of one single molecule, but embraces the general isomerism of a collection of atoms; provided that each atom is used only once, an EM can just as well consist of more than one single species. For example, the EM of the formula $\{C_1H_4O\}$ can have two distinct isomeric appearances: CH_3OH and $CH_2O + H_2$. If EM_1 represents the starting situation and EM_2 represents the terminal situation, the isomerization $EM_1 \rightarrow EM_2$ *is a* chemical reaction. Similarly, a sequence of reactions can be expressed by consecutive isomerizations:

$$EM_1 \rightarrow EM_2 \rightarrow EM_3 \rightarrow EM_4 \rightarrow EM_5 \rightarrow EM_6 \rightarrow \ldots$$

The mathematical representation of the chemical constitution of an EM is associated with a bond-electron (*BE*) matrix. The row/column of a *BE* matrix describes the pattern of valence electrons at the respective atomic core. Nondiagonal entries in a *BE* are formal bond orders between atoms *i* and *j*; diagonal elements correspond to free valence electrons for each atom. The following scheme illustrates the *BE* matrix of formaldehyde. The sum of elements over a row (or column) gives for each atom its number of valence electrons.

$$
BE =
\begin{array}{cccc}
1 & 2 & 3 & 4 \\
\end{array}
\begin{bmatrix}
0 & 1 & 0 & 0 \\
1 & 0 & 2 & 1 \\
0 & 2 & 4 & 0 \\
0 & 1 & 0 & 0 \\
\end{bmatrix}
\begin{array}{c}
H \\
C \\
O \\
H \\
\end{array}
$$

For a given atom type only a limited number of such chemically reasonable patterns (valence schemes) are permissible. The *n* atoms of an EM can be numbered in *n!* different ways, generating an equal number of distinguishable, but chemically equivalent, *BE* matrices. Here, we recall, canonical numbering algorithms are conveniently used to generate one canonical *BE* matrix.[22] Providing for the constancy of the number of electrons during a reaction (neutral EM gives neutral EM, charged EM gives charged EM) and obeying the law of conservation of mass, a matrix **R** can be found such that

$$\mathbf{B} + \mathbf{R} = \mathbf{A} \tag{3}$$

This is the master equation within the proposed mathematical model of constitutional chemistry. **R** is the matrix that transforms a matrix **B** of "educts" (i.e., of molecules before the

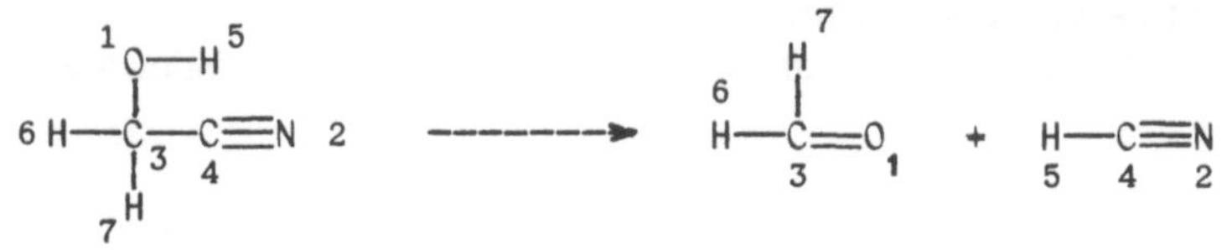

```
   1       5                              7
    O—H                                   H
    |                               6     |
6 H—C——C≡≡N  2      --------→    H—C=═0    +    H—C≡≡N
    |  3  4                          3  1         5  4  2
  7 H                                
```

```
 1 2 3 4 5 6 7        1  2  3  4  5  6  7        1 2 3 4 5 6 7
 _____       _____      _____

 4 0 1 0 1 0 0           +1    -1                4   2
 0 2 0 3 0 0 0                                     2   3
 1 0 0 1 0 1 1       +1       -1                 2         1 1
 0 3 1 0 0 0 0          -1    +1                   3     1
 1 0 0 0 0 0 0       -1    +1                          1
 0 0 1 0 0 0 0                                         1
 0 0 1 0 0 0 0                                         1


      EM          +            R          --------→        EM'
```

FIGURE 5. A detailed representation of the matrix operations for the isomerization of the EM of acetonitrile.

formal reaction) into a matrix **A** of "products" (i.e., of molecules after the formal reaction): **R** is the reaction matrix. Conversely, the following equation also holds:

$$\mathbf{B} = \mathbf{A} - \mathbf{R} \qquad (4)$$

which expresses the reverse reaction. Since the sum of the matrix elements in **A** and **B** must be equal,

$$\Sigma \, (b_{ij}) - \Sigma \, (a_{ij}) = \Sigma \, (r_{ij}) = 0$$

The matrix **R** is symmetric because, due to Equation 3, **B** is a *BE* matrix, and since $b_{ij} = b_{ji}$, $r_{ij} = r_{ji}$. The off-diagonal negative matrix elements $r_{ij} = r_{ji} = -1$ reflect the cleavage of a covalent bond between atoms i and j. Negative diagonal entries r_{ii} indicate the number of free valence electrons lost by atom i during the reaction. Positive off-diagonal elements mark the formation of a covalent bond between the respective atoms. Positive matrix elements on the main diagonal indicate a gain of free valence electrons during the reaction. The reaction matrix **R**, detached from any traditional interpretation attempting to rationalize the driving forces of a chemical reaction, describes a mechanism of formal electron shifting that is able to generate all isomers of an EM.

The scheme shown in Figure 5 proposes the conversion of the EM(hydroxyacetonitrile) into the EM(hydrogen cyanide, formaldehyde) as an example of the formal matrix description of reactions. For simplicity, the zero entries have been omitted in **R** and EM'. This formalism does not necessitate functional groups or substructures previously recognized to perform a chemical reaction. The reaction is seen here as a matrix transformation only. In this context, a dramatic change in the established thinking categories concerning "reaction" and "retro-reaction" seems to be dawning: the dichotomy about what we "feel" to be a real reaction in the forward sense and what we consider just a retrosynthetic, virtual disconnective step (typical of the discussed SDS) suddenly vanishes. Nothing can tell us if the above reaction is a "forward" or a "retro" reaction; only **R** survives.

Exhaustive application of different **R** matrices on a given EM generates all formally possible isomeric EMs; it follows that all possible reaction paths can be generated in principle, the known and the new.

2. The Chemical Distance

An $n \times n$ *BE* matrix **B** also can be represented by a vector **b** with n^2 components:

$$\mathbf{b} = (b_{11}, ..., b_{1n}, b_{21}, ..., b_{n1}, ..., b_{nn})$$

This results in an imbedding of the *BE* matrices of a given FIEM into an n^2-dimensional metric space ω. The matrix elements b_{ij} of **B** can be understood to be Cartesian coordinates of a point P(**B**) in the multidimensional metric space. In the same way another *BE* matrix for an isomeric ensemble **A** corresponds to a vector **a** in ω reaching a point P(**A**). The distance between the two points is the chemical distance[23] and is equivalent to the vector $d_R(\mathbf{A}, \mathbf{B}) = P(\mathbf{A}) - P(\mathbf{B}) = |\mathbf{a} - \mathbf{b}| = |\mathbf{r}|$, but we know that vector **r** in ω corresponds to a *BE* matrix which only can be the reaction matrix **R**, according to Equation 3.

The term "chemical distance" is used to quantify the integer number of valence electrons that must be shifted in order to convert EM(**A**) into EM(**B**). The theory about chemical distance has been illustrated in depth and has interesting aspects concerning the minimal chemical distance accomplishing a chemical conversion.[23] It suffices to say that chemical distance can be related to the ordinary euclidean distance $d_E(\mathbf{A},\mathbf{B}) = [\Sigma (r_{ij})^2]^{1/2}$. It has been shown that the minimum chemical distance $d_R(\mathbf{A},\mathbf{B})$ for a pair of EMs can be found among those matrices $\mathbf{A}' = \mathbf{P}^T\mathbf{A}\mathbf{P}$ for which $d_E(\mathbf{P}^T\mathbf{A}\mathbf{P},\mathbf{B}) = $ minimum, with **P** as the permutation matrix. (The permutation operator is necessary because of the $n!$ ways of numbering the atoms of an EM.)

This result is of relevant practical importance because d_R can only be minimized by trial and error search, whereas d_E can be minimized through the maximization of the scalar product of **A** and $\mathbf{P}^T\mathbf{B}\mathbf{P}$[24] by employing a modified version of a quadratic assignment algorithm used in operations research.[25] The principle of minimal chemical distance says that chemical reactions generally proceed from state to state along a walk of minimal chemical distance, i.e., with a minimal redistribution of valence electrons.

This is itself a method for the optimization of the evaluation strategy of reaction paths in computer-assisted reaction simulation, referred to as the bilateral solution of multistep syntheses. With two given *BE* matrices, **A** and **B**, the decomposition of the **R** matrix $\mathbf{R} = \mathbf{A} - \mathbf{B}$ into the components $\mathbf{R}_1, \mathbf{R}_2, ..., \mathbf{R}_r$ such that

$$\mathbf{R} = \mathbf{R}_1 + \mathbf{R}_2 + ... + \mathbf{R}_r$$

gives the data on the chemical pathways which account for the overall transformation. The network of reactions departing from a certain **T** therefore can be controlled at the user's will and beamed toward known structures **A** (members of a library of educts, for example) in a retrosynthetic study. The reaction intermediates can be ranked according to their strategic usefulness by computing their actual chemical distances from the path of minimal chemical distance. This strategy is obviously conceivable if one accepts that paths of minimal distance are really the best walks along the synthesis tree, a statement which is not always true. Sometimes slight detours can result in reaction steps that are easier to perform in practice. Other situations with known final states EM(**A**) are found in biochemical reactions. The attempts at elucidating metabolic pathways are facilitated by the computerized processing of the substrate and of the metabolite EMs; this is due to a partitioning of the global transformation $\mathbf{B} \to \mathbf{A}$ into elementary, single-reaction steps. The crop of a formal treatment of organic reactions is primarily the abandonment of the concept of directionality of a chemical reaction. Using this formal treatment, programs have both synthesis design and reaction prediction capabilities at the same time. Just the physicochemical evaluation of the single reaction steps and some higher strategies supported by heuristic information make the difference.

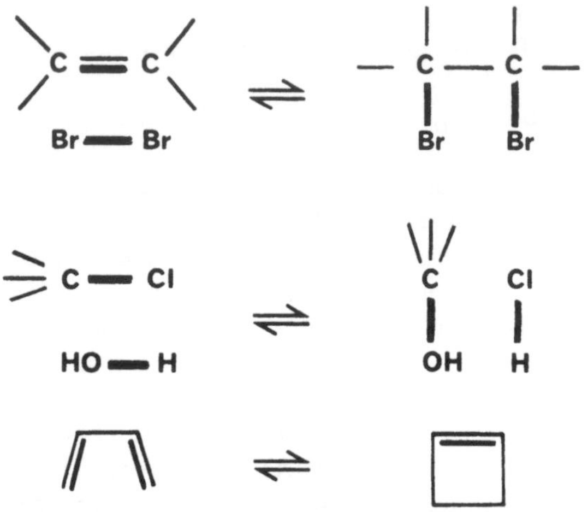

FIGURE 6. Different reaction classes, all of which have one important feature in common. (Which one?)

B. REACTION GENERATORS

The completeness of an algorithm that applies all possible reaction matrices to a certain EM reveals the weak side of a formal approach to the modeling of chemical reactions. Theoretically, all states of a problem space around an EM can be generated, leading to a combinatorial explosion. If all bonds in a certain EM were sequentially ruptured in all possible ways and new ones were formed, the number of generated isomers and the corresponding chemical reaction would become too large to be managed. Worse yet, an enormous number of impossible, chemically unrealistic reactions would be produced. If not kept under intelligent control, the system turns out to be just a bond-crunching mill.

We must therefore ask which reaction matrices are really necessary to model the apparent wealth of possibilities in organic chemistry, and then we must understand when to apply them.

To fully comprehend the far-reaching importance of **R** the following argument might be of help. Consider a number of students being taught organic chemistry. Half of them use a modified textbook about organic chemistry in which the pages about, say, electrocyclic reactions are omitted intentionally, while the other half use an uncut version. Thus, half of the students will receive a chemical education without having ever seen an electrocyclic reaction equation. Will someone in that group be able to conceive a reaction of this type, unknown to him, within a reasonably short period of time? Someone probably will develop such a reaction scheme, but no one can tell in advance when the finding will occur.

The following classes of reactions are usually treated in different chapters of standard organic chemistry textbooks: addition and elimination reactions, substitution reactions, and electrocyclic reactions. They are often several hundred pages apart. However, in chemical terms, are they really so far apart?

Before reading further, the careful reader should evaluate the reaction schemes of Figure 6, corresponding to the mentioned classes, and should try to detect what they have in common. The depicted examples representing three different classes of chemical reactions (that is the orthodox way in which they are presented to students) can, if we look at chemistry from a completely alternative side, be considered as belonging to one single class: the class of reactions involving the cleavage of two bonds and the formation of two others. Thus, all three reactions can be described by just one type of **R**, containing two -1 and two $+1$ entries. We call such a matrix in its compressed form a reaction generator (RG). The RG responsible for breaking any two bonds (I–J and K–L) and joining two others (I–K and J–L)

is an RG22 generator. We now can conclude the psychological experiment with the students: if one of them knew the RG22 formalism, he certainly would find the bond-shifting scheme leading straight to what we call an "electrocyclic reaction".

We must understand that in the world of RGs the meaning of novel and known reactions is lost. Potentially they are all contained in the operator RG. In an exhaustive way, on a given EM, its application gives access to all formally possible reactions (and products), i.e., to all isomeric EMs. The computer-assisted modeling of organic reactions (retrosynthetically or in a forward sense) in principle can be achieved without having to represent individual reactions stored in a reaction library. By gaining access to all conceivable reactions through the RG formalism the foundation is laid for the discovery of new reactions.

Systems working with such formal representations of bond and electron rearrangements have been developed with varying degrees of sophistication (AHMOS,[26] ASSOR,[27] CAMEO,[28] IGOR,[29] EROS[30-32]). EROS is geared completely on five kinds of RGs. It may seem astonishing, but 99% of known organic chemistry can be modeled by the five reaction generators shown in Figure 7.

RG12 breaks one bond and forms two others; RG21 breaks two bonds, forming a new one; RG33 breaks three bonds, joining three new ones; and RG221 models the dipolar cycloadditions, cleaving two bonds, making two, and shifting one free electron pair.

Once again, reaction schemes normally described in different chapters are united under a unique RG mechanistic representation like, for example, the Cope and the Favorski rearrangements: they are both modeled by an RG33 operator. Also, a Diels-Alder reaction is reproduced simply by an RG33 operator, which breaks three bonds and forms three new ones.

These particular formal features of the SDS EROS make it very attractive to the researcher who is seeking novelty.

Scheme 1 illustrates how a retro-aldol condensation can be represented in EROS by manipulation of the *BM* matrix. The bond order of the atom pairs involved in the chosen substructure is lowered by one, and the new bond relations (1–2, 5–6) are stored in a separate matrix. The latter is added to the original *BM* matrix after deletion of all columns where $BM(i,3) = 0$. Since the procedure treats atoms and bonds as formal elements, a recombination like

$$H - \overset{\displaystyle \diagdown}{\underset{\displaystyle \diagup}{C}} - O - C \equiv$$

is also conceivable. It would correspond to the bond-making pattern *I–L* and *K–J*. However, this solution is chemically much less significant than the aldol condensation.

If left autarchic in the application of the RG operators, the system clearly suffers from a nonnegligible weakness: it does not discern good applications of the available RGs from bad ones. It is now up to the computer chemist to develop selection strategies and tactics to extract from the bulk of formal possibilities those which have chemical significance.

C. EVALUATION TACTICS IN EROS

The evaluation of reaction routes in a retrosynthetic study and of the reactivity of bonds in a forward-search run is achieved in EROS by inclusion of heuristic rules and physico-chemical parameters. Due to the different schemes of "reasoning" that the computer adopts in synthesis planning and reaction prediction, it seems adequate to separate their discussion into two subsections, starting with synthesis design.

Synthesis design has many features in common with a chess game. To compare them let us analyze the following correspondences:

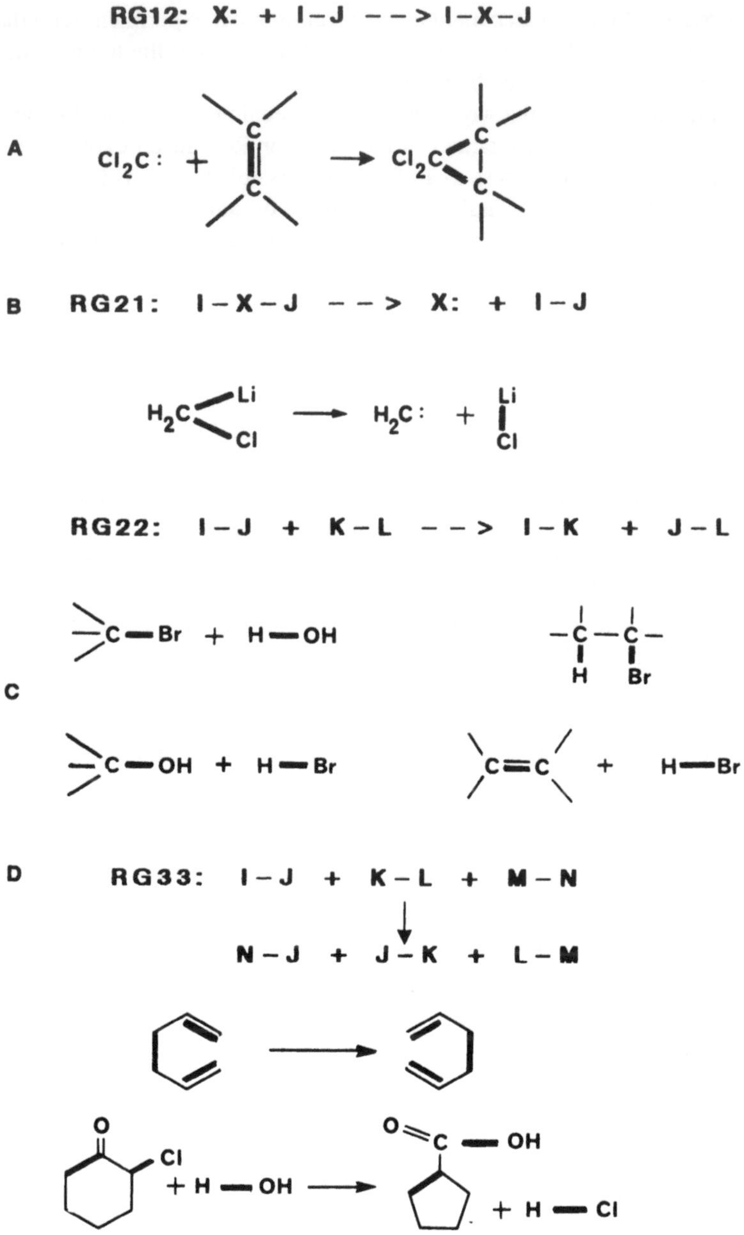

RG12: X: + I–J – – > I–X–J

A

B   RG21:   I – X – J   – – >   X: +   I – J

RG22:   I – J   +   K – L   – – >   I – K   +   J – L

C

D   RG33:   I – J   +   K – L   +   M – N

N – J   +   J – K   +   L – M

FIGURE 7.   An overview, with examples, of the five reaction generators working in EROS (A, B, C, D, and E).

Feature	Chess Game	Synthesis design
Objects	32 pieces	Millions of compounds
Rules	Moves, defined exactly	Reactions, no general theory of chemical reactivity
Planning	Always forward	Backward
Goal	Checkmate	All starting materials suited for the synthesis (cost, availability, etc.)

The compilation of this table shows that chemical problems are many degrees more complicated than chess games. First, in organic chemistry there are many more objects to consider

E   RG221:   I—J + K—L + X: --> I—K + L—X + J:

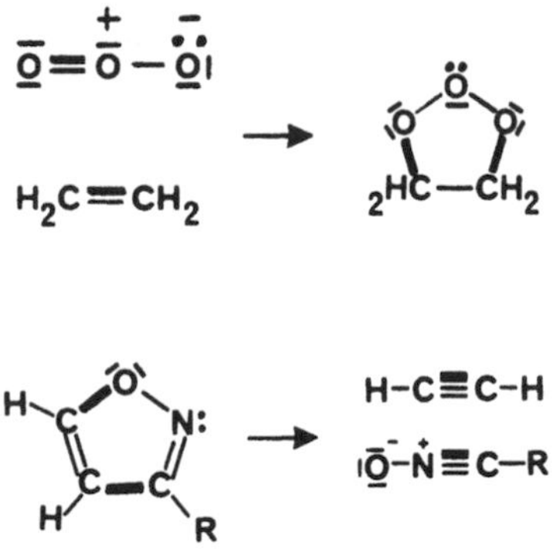

FIGURE 7E.

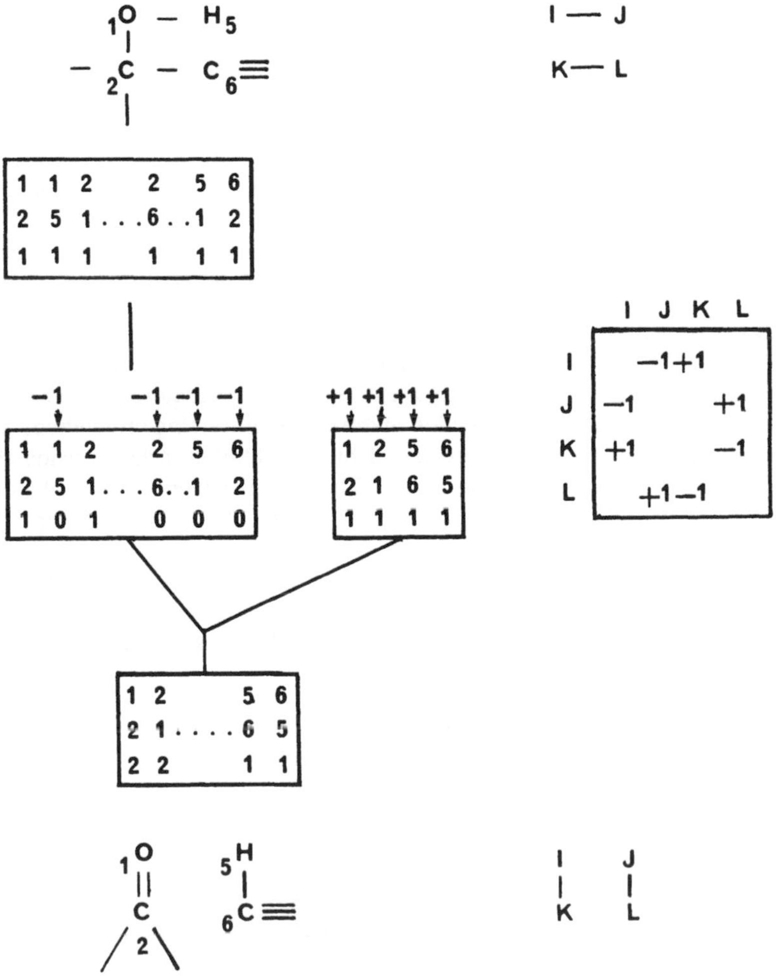

SCHEME 1.   The formal *BM* matrix description of a retro-aldol condensation.

than the 32 pieces on a chess board. Chess rules are well defined and have no exceptions, whereas chemistry rules, according to which we all plan our syntheses, are distinctly more vague, with sudden pitfalls, irrational irreproducibilities, and empirical rules-of-thumb. Given a certain target, there is no generally valid prescription for finding the best synthesis route. Nor can we rely on a sound theory of chemical reactivity to predict products from a set of reactants with certainty.

Nevertheless, any SDS modeling organic chemistry in a global manner has to address the problem of predicting chemical reactivity. For example, in the simple reaction scheme

$$I\text{--}J + K\text{--}L \left\{ \begin{array}{ll} \longrightarrow I\text{--}K + J\text{--}L & \text{(a)} \\ \longrightarrow I\text{--}L + J\text{--}K & \text{(b)} \end{array} \right.$$

the ease of reaction (a) prevailing over reaction (b) (and vice versa) will be determined by factors such as

- Identity of atoms I, J, K, and L
- Bond order between these atoms
- Vicinal and remote neighborhoods of these atoms
- Reaction conditions

EROS has modules describing these factors in physicochemical terms as much as possible, using values for bond energies,[33] atomic charges, residual electronegativities, bond polarizabilities, etc. To calculate these values, various models and algorithms (e.g., the models *PEOE* and *SD-POE*, presented in Chapter 6, which are used to calculate charge distributions) have been included in EROS. Mathematical expressions are set which use these values to establish the quality of predicted reaction pathways. The favorable sequences of reactions are selected according to these evaluation tactics.

As an example, the calculation of one parameter strongly governing the destiny of reacting molecules, the reaction enthalpy, is discussed in more detail.

1. Evaluation of Reaction Enthalpies

Heats of reaction[33] are calculated automatically in EROS via an algorithm that uses thermochemical parameters.[34] The algorithm is extremely fast, as required by an SDS because every retrosynthetic run or reaction prediction involves the processing of hundreds of bonds. EROS solves the problem with a method based on an additivity scheme using the well-known topological representation of molecules. The additive thermochemical parameters refer to the atomization enthalpies for bonds between specific atoms, the sum of which yields an estimate for the formation enthalpy of a molecule.

For example, for isobutane we have three C–C bonds (C_1–C_2, C_2–C_3, and C_2–C_4), ten C–H bonds, three C–C–C units (C_1–C_2–C_3, C_1–C_2–C_4, and C_4–C_2–C_3), and one unit with a central carbon atom (C_2) linked to three carbons. These elementary units are parameterized with specific enthalpic constants B, G, and D.

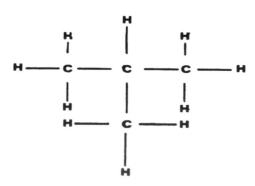

The B parameters refer to atoms in C–C and C–H bonds (80.96 and 98.97 kcal/mol, respectively); the G parameters refer to atoms of carbon-carbon bonds inside of a chain of three carbon atoms C–C–C (1.1 kcal/mol); the D parameters belong to a C–C bond departing from a central carbon atom surrounded by three other carbons (-0.06 kcal/mol). The formation enthalpy H^0 can be computed for isobutane by the equation

$$H^0 = 3B_{CC} + 10B_{CH} + 3G + 1D = 1235.82 \text{ kcal/mol}$$

The reaction enthalpy is given by

$$\Delta H_R = H(educts) - H(products) \tag{5a}$$

A remarkable optimization of the algorithm is introduced, considering that in a reaction simulated by EROS everything remains constant except the bonds processed by a given RG operator. Equation 5a therefore contains a varying excess of useless information, namely, all the parameterizations for untouched bonds. It suffices, in fact, to take into account only those atoms around which the reaction occurs. We can substitute Equation 5a with Equation 5b,

$$\Delta H_R = H(bonds\ broken) - H(bonds\ made) \tag{5b}$$

Numerically, one obtains the following for the conversion butane $\rightarrow$ isobutane with Equation 5b:

$$\text{CH}_3\text{--CH--CH}_2\text{--CH}_3 \rightarrow \text{CH}_3\text{ --CH--CH}_3$$
$$\qquad | \qquad\qquad\qquad\qquad |$$
$$\qquad \text{H} \qquad\qquad\qquad\qquad \text{H--CH}_2$$

$$\Delta H_R = (1B_{CH} + 1G + 1B_{CC} + 1G) - (1B_{CH} + 1B_{CC} + 3G + 1D)$$

$$\Delta H_R = -(1G + 1D)$$

The computation of ΔH_R now requires only 10 parameters instead of 32 if only atoms I, J, K, and L are evaluated.

A characteristic feature of the algorithm is that the computation times are independent of the size of the processed molecule. The program only has to perform a topological search for the B, G, and D parameters inside a constant boundary of bonds determined by the actual active RG. The previously presented isomerization of butane can be condensed into two substructures which require the same ten thermochemical parameters for the butane $\rightarrow$ isobutane conversion as for, say, undecane $\rightarrow$ isodecane or for any other RG22-driven reaction for nonstrained alkanes. EROS has a compact library of thermochemical parameters for the most frequent bond types. The predicted ΔH_R values are close enough to the experimental data to guarantee a realistic reproduction of the enthalpic side of a chemical reaction.

Several enthalpies for a bromination reaction of alkanes have been calculated with the EROS algorithm and provide us with experimental values as an illustrative example. The results are quite encouraging, and they allow the computation of thermochemical data for still unmeasured molecules.

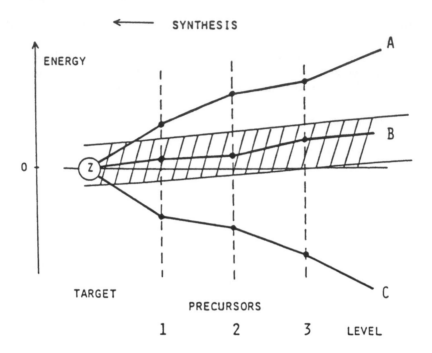

FIGURE 8.   The energetic window. Optimal retrosyntheses are to be searched inside the shaded channel.

ALKANE	+	BROMINE	→	BROMOALKANE;	ΔH_{EXP}	ΔH_{CALC}
Methane			→	Bromomethane	−7.24	−8.41
Ethane			→	Bromoethane	−11.10	−11.08
Propane			→	1-Bromopropane	−11.41	−11.08
Propane			→	2-Bromopropane	−14.42	−13.93
Butane			→	1-Bromobutane	−11.51	−11.08
Isobutane			→	2-Bromoisobutane	−15.75	−16.96
Pentane			→	1-Bromopentane	−11.72	−11.08
Pentane			→	2-Bromopentane	?	−13.93

Note: Enthalpy units kcal/mol.

In a forward search the reaction enthalpy can be used directly to rank the probability of a set of predicted reactions; for example, the bromination of propane leads to 1- and 2-bromopropane. The latter reaction is predicted to be much more exothermic than the former. Assuming thermodynamic product control (as EROS does in this situation), the calculated enthalpies already correctly rank the two competing reactions. Reality shows that 1-bromopropane is only a by-product of the other isomer.

In a retrosynthetic study the calculated enthalpy value will be used with a different sign and emphasis. This is illustrated with the aid of the energy diagram shown in Figure 8.

In synthesis design it is not advantageous to select exothermic retroreactions because the synthesis itself is performed in a reverse direction to the simulated reaction route. Thus, exothermic retroreactions (as in path C of the energy diagram) encountered in the synthetic direction would demand an input of energy into the system, a situation often unfavorable from an economic and tactical point of view. On the other hand, a search along path A leads to the prediction of precursors and of starting products which are so high in energy that they are unmanageable and unstable for practical synthetic work. For these reasons it is advisable to search the problem space along paths of type B, which are located inside a certain enthalpic range, the energy window, represented by the hatched area in the diagram. Within the window all simulated retroreactions are slightly endothermic, resulting in a smooth, controllable exothermic character in the synthetic process. Thus, at the start of an EROS run the user can decide about the width of the energetic window and thereby control the inclusion or exclusion or reactions at the window's border. This option, among others influencing the final shape of the synthesis tree, can be used effectively in steering the simulation toward routes more appropriate for laboratory synthesis or for industrial processes. Industrial, large-scale processes are syntheses involving only a few steps and can use more extreme reaction conditions (e.g., higher temperatures, pressure). For these kinds of processes the window can be widened and more endothermic reactions allowed in the simulation course.

2. The SOHIO Process Discovered with EROS

For example, in the SOHIO process[35] (Figure 9) for producing acrylonitrile, water is formed as a by-product. In an EROS retrosynthetic run, water later was added to the target, acrylonitrile. There are two ways of adding water to the triple bond of the nitrile group. In this case, the more endothermic route has been chosen because the synthesis itself is then more exothermic and, consequently, very favorable. The value of 32.9 kcal/mol already would be a quite high value for ordinary laboratory reactions. This is not true in an industrial environment, where higher energies and different equipment are available. The addition of a second water molecule is again an endothermic step, and the third and last step is exothermic, rebalancing the overall energy score. We see that three reaction steps are sufficient to lead from the target, acrylonitrile, to propene and nitrous acid. However, this is not the SOHIO process chemistry (which can be modeled by selecting the first, less endothermic route). Nevertheless, an interesting idea is revealed by EROS in substituting ammonia with nitrogen in some oxidized form (nitrous acid, nitrogen oxide), a method which could be used to convert propene into acrylonitrile. Indeed, there is a process based on utilizing NO: it is the DuPont process.[37] Had this alternative path not been known before, this study would have suggested trying an alternative approach to the SOHIO method.

It should be clear that all three reaction steps of the above synthesis tree were generated autodeductively by the SDS EROS with just the RG22 operator. This simulation did not require the special treatment customary for an SDS working with a library of reactions.

3. Retrosynthesis of a Prostaglandin-Like Compound

Another study shall be discussed here briefly to exemplify the use of EROS. A target compound having a prostaglandin-like structure delivered several first-level precursors of varying strategic relevance. Some of them are shown in the collection of Figure 10.

EROS suggests ideas. Functional groups sometimes may not be understood literally, but eventually must be refunctionalized in the researcher's mind. Reagents, protecting groups, and solvents must be added by the user in the interpretation of the results. Therefore, emphasis is given more to the architecture of a disconnective step than to local particularities. The target was processed with one water molecule as coproduct.

Precursor 10361 yields the target with an exothermic reaction of −12.5 kcal/mol. The enolic substructure is reformulated immediately as an aldehyde and the methyl group functionalized by a Grignard functone. Further dissection leads to precursors 2 and 3, the former conveniently further simplified into starting materials 4 and 5 (Figure 11).

Precursor 10101 can be dissected by a retro-aldol into compounds 6 and 7, both having α-acidic hydrogens. (Problems of self-condensation can arise here in practice.) Precursor 6 can be synthesized by methylation of ester 8 via organometallic reagents (Figure 12).

Precursor 10071 can be retrosynthesized by cleavage of the double bond and yields precursor 9, which possibly might be obtained by double condensation of 10 and 6 (Figure 13).

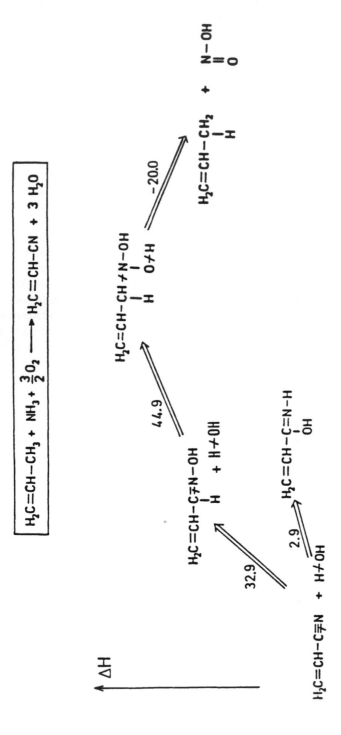

FIGURE 9.   The SOHIO process as formulated by EROS.

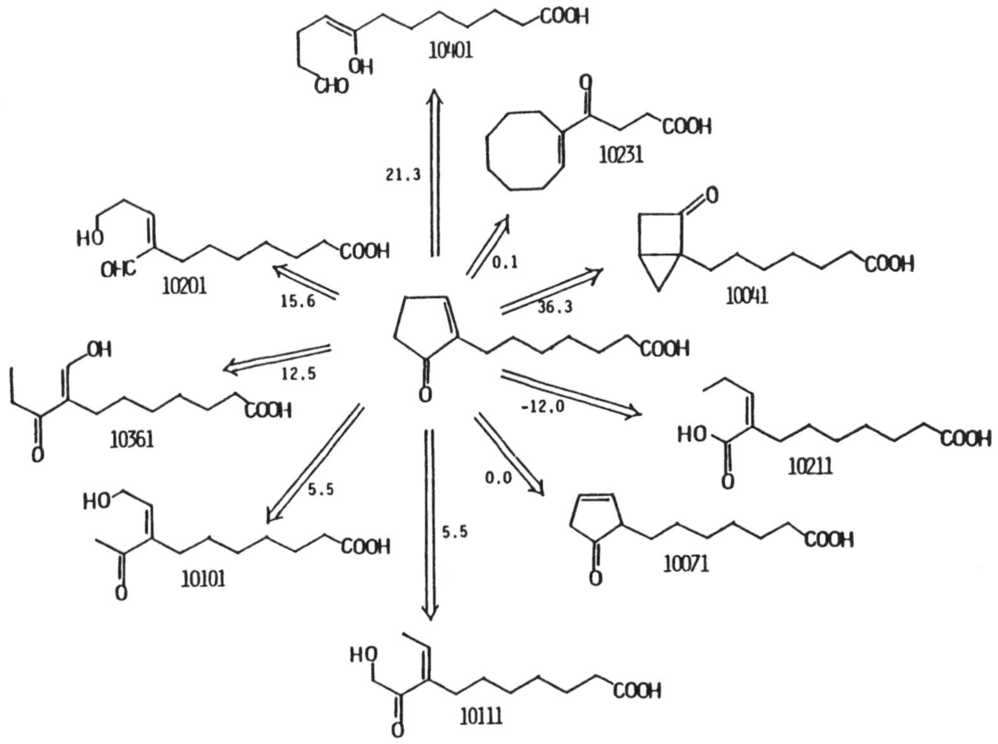

FIGURE 10. A selected number of first-level precursors generated by EROS in a retrosynthesis of the target PGE-like compound. Numbers marking the arrows refer to the predicted retroreaction enthalpies.

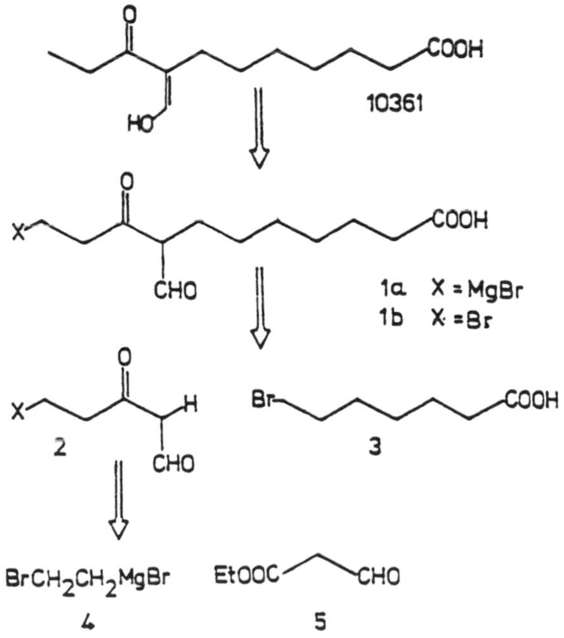

FIGURE 11. One possible synthesis of the target from precursor 10361.

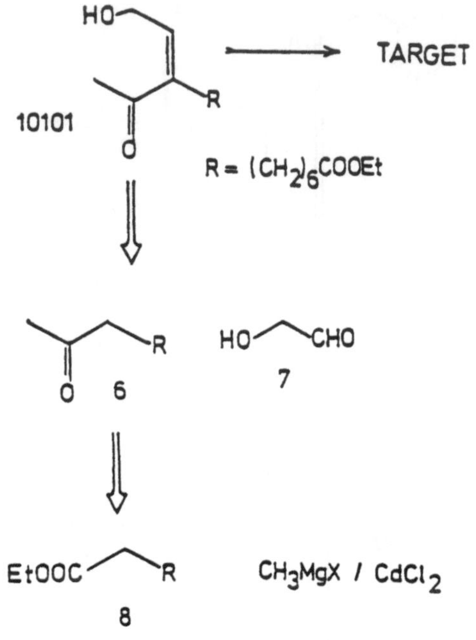
HO
TARGET
R
10101
O
R = (CH2)6COOEt
R
O 6
HO CHO
7
EtOOC R
8
CH3MgX / CdCl2

FIGURE 12. Another path leading to the PGE analogue via precursors 6, 7, and 10101.

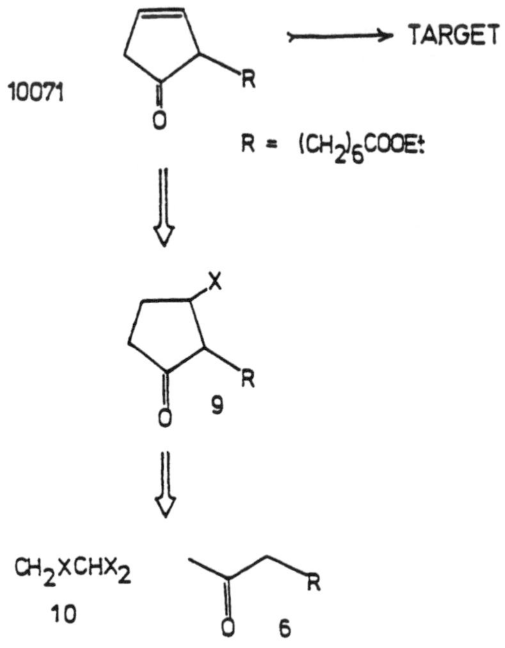
TARGET
10071
R
O
R = (CH2)6COOEt
X
R
O 9
CH2XCHX2
10
R
O 6

FIGURE 13. Another possibility for reaching the target is offered by precursor 10071, the synthesis of which is conceivable through condensation of compounds 10 and 6.

Other tactical inferences for the evaluation of synthesis paths are given in EROS by ring strain energies[37] and delocalization energies.[39]

The reaction enthalpy is an important (but not the only) parameter determining the course of a simulated (or real) reaction. Other factors are of an electronic nature and find their application especially in the forward prediction of reactions. The concept of "chemical reactivity" recovers its original meaning, obscured by the strange "backward reactivity" encountered in retrosynthesis. When we speak of reactivity, in the majority of cases we want to know which bond in a molecule reacts first in a given environment. The computerized modeling of chemical reactivity must be tailored around some model for the automated evaluation of the reactivity of bonds.

IV. CHEMICAL REACTIVITY AND FORWARD SEARCH

A. ELECTRONIC EFFECTS

The quantitative treatment of electronic effects and of their relative role in the prediction of reactions is a very difficult problem, and only partial (although promising) solutions have been reached so far. Basic electrostatic considerations indicate that the cleavage of a covalent bond into formal positive and formal negative residuals is intrinsically an endothermic process. It will be facilitated by all intervening mechanisms providing charge dissipation. Chemists are mentally trained to rationalize electronic stabilization according to various effects like partial charges, electronegativity, inductive and resonance effects, polarizability, hyperconjugation, hydrogen bonding, and others. Sometimes, if not always, a number of the listed "effects" act simultaneously, making a clearcut distinction of stabilizing mechanisms and, consequently, their paremeterization in quantitative terms really arduous.

If, on one hand, the reaction enthalpy evaluation procedure takes into account only the ground states of the educts and products, the role of the electronic evaluation parameters seems to be heavily dependent on the transition state of the reactants. Take, for example, the simplest but probably one of the most fundamental reactions in organic chemistry: the abstraction of a proton from a substrate. Intuitively we assume that the more positive the partial charge of the hydrogen atom, the more likely it will be detached from the molecule. This turns out to be only a partial truth.

An RG22 operator has, for example, several formal ways of heterolytically breaking a bond containing a hydrogen atom in the small molecule of Figure 14. Depending on both the solvent and the abstracting base, a few of the possible formal mechanistic steps may model reality, but others are just unrealistic, formal bond ruptures. How can an autodeductive program like EROS discern between applicable and unreasonable heterolytic bond cleavages? How can it detect the resulting polarity of the fragments and eventually rank breaking priorities?

In Chapter 4 we introduced the notion of partial atomic charges calculated from a fast empirical model relying on orbital electronegativity. The atomic charge as such is just a ground-state property of the molecular system and cannot reasonably describe the reactivity of bonds when major charge rearrangements (of an inductive or mesomeric kind) occur in the reacting species. Charges were correlated with reactivity parameters, such as the pK_A values of the hydrides of the atoms from the first long period,[39] yielding a fair correlation; if molecules like HCN or ethylene were included, the correlation worsened. Similarly, the correlation of the charge at the carbonylic carbon of substituted acetic acid esters with the σ^* Taft constants[40] of nucleophilic displacement reactions does not show a significant covariance between the two parameters.[41] In general it was found that a one-variable equation was not sufficient, using simple empirical models, to model the reactive behavior of chemical bonds sufficiently well to be used for simulation and prediction purposes.

According to what we said before, it is mandatory for a reaction modeling program to

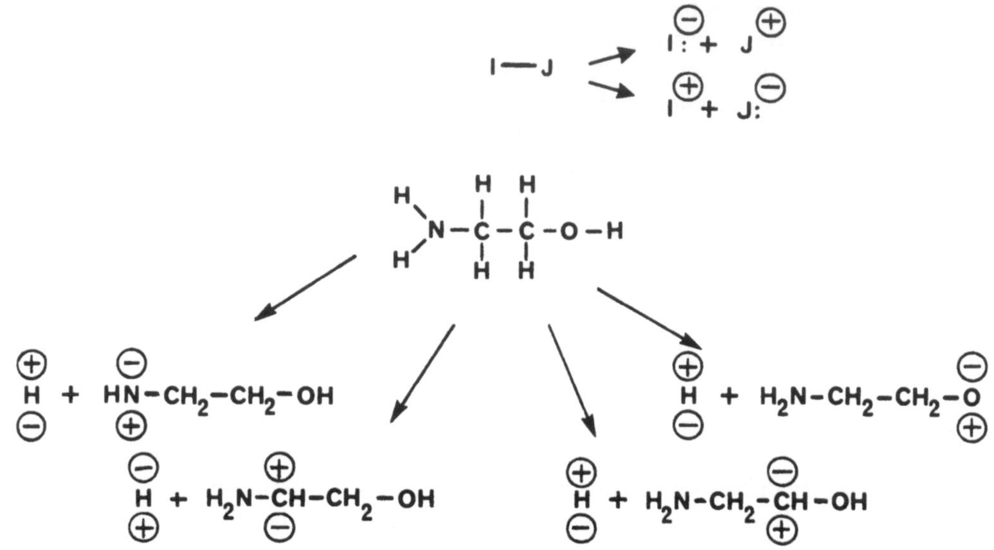

FIGURE 14. Formal ways of breaking bonds containing a hydrogen atom, as obtainable with an RG22 operator. Some are realistic, and others are energetically highly improbable.

evaluate the required parameters in the shortest possible CPU time. The "description" of a transition state therefore must be accomplished in an empirical manner, like the computation of the other reaction-determining molecular parameters. The characteristic features that are known to stabilize a transition state, thus favoring the reaction process, must be approximated from ground-state parameters.

It was found that the effective polarizability and the residual electronegativity (see Chapter 4) were revealing in the attempt to model reactivity, reproducing in a simple but effective way the charge reorganization in a (mechanistic) transition state. The residual electronegativity χ_R is the particular final *OE* value reached at the convergence of a charge computation according to the algorithm based on orbital electronegativity, which was explained in detail previously. Since the residual electronegativities of the atoms in a molecule are not equalized, there is still a latent possibility of charge shifting between the atoms provided that a perturbation acts on the system. Stated differently, the residual electronegativity is a measure of how an incoming perturbation at the electronic level can be "absorbed" and stabilized. This parameter gives us a possibility of predicting what would happen in a real transition state just from considering a ground-state property (e.g., the residual electronegativity), which is offered at no further expense by a standard charge calculation.

If the simulation program had to decide about the relative reactivity of two molecules like X–CO–OR and Y–CO–OR in an esterolytic reaction, for example, the following two possibilities would arise:

$$NUC^- + X\text{–}CO\text{–}OR \rightarrow NUC\text{–}CO\text{–}X + RO^-$$

$$NUC^- + Y\text{–}CO\text{–}OR \rightarrow NUC\text{–}CO\text{–}Y + RO^-$$

If we assume that the bond dissociation energies are approximately equal for the C–OR bond in both substrates, the reactions would be judged as equally probable (i.e., the mentioned bond would be of comparable reactivity in both systems).

Qualitatively, for known substituents and within a standard molecular reference frame, the inductive Taft substituent constants σ^* can be used to rank reactivities. For example, a charge calculation gives for the carbonylic carbon in $Cl_3C-CO-OR$ a value of $+258$ millielectrons, and $+242$ millielectrons are attributed to the equivalent atom in $HOCH_2-CO-OR$. There is only a negligible difference in charge between these two atoms, but on the Taft scale the discrepancy amounts to an amazing 2.1 units.

Because NUC and OR are kept constant, the reason for this marked difference in (experimental) reactivity must be located within the X and Y groups, and it is precisely a combination of the stabilizing effects of the polarizability and of the residual electronegativity (of the first and second neighbor spheres around the reacting atom) that accounts for the experimental evidence.

Using functions of the kind

$$y = c_0 + c_1\chi_R + c_2\alpha_D \qquad (6)$$

it became possible to model proton acidity,[42,43] Taft substituent constants,[41] and proton affinities for alkylamines[44] as well as for alcohols, ethers, thiols, and thioethers.[45] These functions establish an encouraging attempt to design reactivity models without having to enter the realm of quantum chemical simulations.

Simple linear equations could also be developed for other systems: the proton affinity of aldehydes and ketones, and their hydride ion affinities.[46] However, in addition to effective polarizability and electronegativity, hyperconjugation has also been used as modeling parameter, since p orbitals carrying positive charge are involved in the reactions.

These investigations were aimed at confirming a prototype model for reproducing chemical reactivity, parameterizing a key (but nevertheless mechanistically quite simple) family of reactions in the gas phase. In many organic reactions the situation is far more complex and involves more than just a proton (or hydride) transfer between substrates of one and the same class. For example, if a reaction generator RG22 acts on a molecule, it must dissect two bonds and join two others. If, by means of the above empirical equations, the first bond (the most reactive one) has been determined and heterolytically ruptured, two formal charges appear in the molecular frame. The generated charges inductively influence the whole skeletal neighborhood in a specific manner, largely modifying the reactivity of the surrounding bonds. Thus, a second evaluation of bond reactivity must be performed in order to detect which bond turns out to be the most reactive depending on the identity of the first broken bond. If the RG operators could act simultaneously on two bonds ($I-J$ and $K-L$) in retrosynthesis, this is no longer acceptable in forward search, as an asynchronous bond evaluation seems to reproduce experimental reality more realistically. The method consists of splitting a reaction **R** into two half-reactions **R1** and **R2**, where **R1** is the half-reaction cleaving bonds and **R2** is the one joining bonds. After selection of a first breakable bond (identifiable through a reactivity function similar to Equation 6) the reaction generator breaks it heterolytically, choosing the polarity so as to provide the highest amount of stabilization in the following manner: by applying a charge calculation to the intermediate or "transition-like" state of the processed molecule, i.e., the state generated by the action of **R1** on the first bond, one obtains an empirical measure of the displaced charge in the transition state (within the EROS mechanistic representation of a reaction). The better the charge created upon heterolytic bond rupture is spread over the residual molecular framework, the higher the probability that this particular bond cleavage occurs. The bond cleavage now can take place with the right polarity, the excess formal charges giving rise to a charge redistribution process which is governed mainly by the residual electronegativity of the other atoms. To determine the second bond in the pair of bonds treated by RG22, the program performs an analog search for the next most breakable bond, processing the perturbed molecular system as just described.

It could be shown that the same parameters so successful in quantifying gas-phase

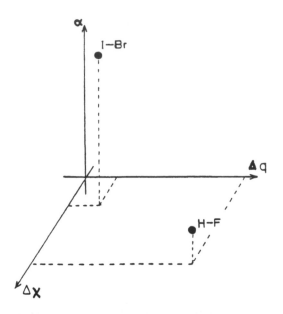

FIGURE 15.   The reactivity space spanned by three variables: the polarizability of bonds, the difference in residual electronegativity of the atoms in a bond, and the difference in charge between the two atoms along this bond.

reactivity data are also applicable to reactions in solution (and here we come nearer to our traditional image of wet chemistry). The reactions studied included aqueous-phase acidity of alcohols and gem-diols as well as hydration of carbonyl compounds.[47]

B. THE REACTIVITY SPACE

The real complication in modeling chemical reactions is that they are influenced by many factors simultaneously and to various and changing extents. To account for the dependence of reactivity on many variables (α_D and χ_R are insufficient for the bulk of different chemical bonds), a conceptual extension into a multidimensional parametric space was introduced. This was also the consequence of less accurate and fewer reactivity data available for other classes of compounds.

In the multivariate approach the parameters calculated by the different methods are taken as coordinates (of basis vectors) in a multidimensional reactivity space.[48] A bond of a molecule is represented by a vector in such a space, the elements of which are the numerical values of each parameter. For example, as shown in Figure 15, in a reactivity space spanned by the basis vectors α_D, Δq, and $\Delta\chi_R$ the bond in iodine bromide is distinguished by a high polarizability, but small differences in charge and residual electronegativity. The opposite is true for the bond between hydrogen and fluorine. In a similar manner the various bonds of an organic compound will each be represented by a point in the reactivity space. In practice the space is seven-dimensional in the EROS program, involving the following quantities as basis vectors:

- Bond dissociation energies
- σ charges
- π charges
- Inductive effect via χ_R
- Resonance effects
- Hyperconjugation
- Effective polarizability α_D

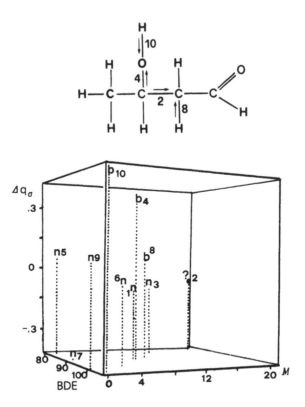

FIGURE 16. The reactivity space, showing the distribution of
selected bonds of the aldol molecule. Reactive bonds (breakable)
are marked *b*, nonreactive bonds (nonbreakable) are labeled *n*.
Arrows show the polarity of heterolytic bond cleavage.

Such spaces are a suitable tool to further understand to what extent the various chemical
effects influence the definition of "reactivity" of organic compounds. Since polar processes
are of outstanding importance in organic chemistry, heterolytic bond cleavage was studied
preferentially. The quantitative treatment of the homolysis of bonds is included in the
mathematical formalism of the former model and represents a border case, with the bond
dissociation energy being the prevailing factor.

When considering the heterolytic cleavage of bonds, each bond will be represented by
two vectors corresponding to the two possible polarities in the fragments. Figure 16 shows
several single bonds of acetaldol in a reactivity space spanned by the polarity in the σ electron
distribution, $\Delta q(\sigma)$, the bond dissociation energy (BDE), and the resonance effect M (given
by the amount of stabilizing π charge generated by the breaking of polar bonds).

Calculations were performed for each bond of the molecule by the presented empirical
methods to obtain the values for the three vector components. The two possible patterns of
bond heterolysis are illustrated graphically with points 4 and 7. Point 4 corresponds to the
dissociation of OH^- from the main molecular frame, whereas point 7 represents the loss of
an OH^+ fragment.

The bonds were labeled according to reactivity. Bonds that were considered by the
chemist to be reactive (breakable) are marked with a *b*, while those found to be inert are
labeled with an *n*. The central C–C bond was left unclassified. The C=C double bond is
not represented here, since the study is devoted to understanding the reactivity of single
bonds.

The picture shows that reactive and nonreactive bonds are separated (clustered) in the

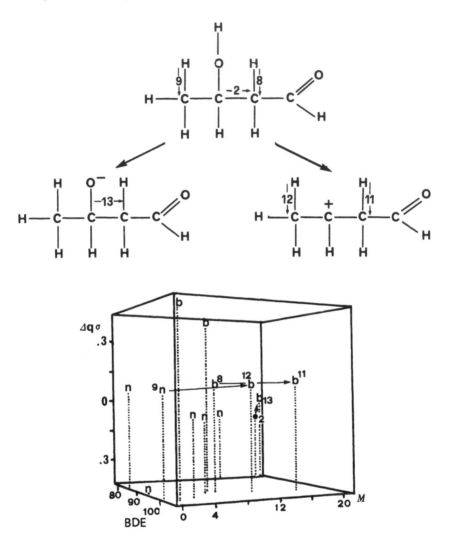

FIGURE 17. Upon deprotonation of aldol or after abstraction of a hydroxylic anion, particular bonds that were nonreactive in the neutral state become breakable. Their positions are shifted into regions of higher bond reactivity inside the reactivity space.

reactivity space. The separation is a confirmation of a method that allows reliable discrimination between breakable and nonbreakable bonds in a multiparametric reactivity space. It is tempting to assume that the more reactive a bond is, the further the point representing it should be away from the plane of separation. From the perspective in Figure 16, the more reactive bonds should be localizable in the region of the upper right and rear corner. This expectation coincides with reality and is illustrated in Figure 17. This diagram shows some additional points for bonds corresponding to two ions obtainable from the aldol. Computations of bond dissociation energies, σ charge distributions, and π charge stabilization through resonance were carried out for all bonds of the aldol anion and for the carbocation obtained from it after removing an OH^- group. Only three of those bonds are visualized in the reactivity space for clarity.

It is well accepted that the central C–C bond of aldol becomes more reactive after deprotonation allowing a retro-aldol condensation. This increase in reactivity is reflected in the shift of point 2 to point 13, away from the observer in a direction of higher reactivity, as postulated before. The removal of a proton from carbon atom 2 of aldol leading to the enolate anion, represented by point 8, was regarded as a possible event. In the carbocation, removal of this proton is given by point 11. This situation must be facilitated, since the incipient carbanion will be stabilized by the adjacent positive center, yielding an α,β-unsaturated carbonyl species. This anticipated increase in reactivity is reflected in the shift of point 8 to point 11, into a region of higher reactivity.

Removal of a proton from carbon atom 4 of aldol was not considered to be easy (point 9). In the carbocation, however, this proton should be much more acidic: the resulting carbanion can be stabilized by the vicinal positive charge. Therefore, this bond becomes reactive, and the representing point for third heterolysis is relocated more toward the right side of the reactivity space. However, the product is a β-unsaturated carbonyl compound, which is not as stable as an α,β-isomer. This is also visible from the fact that point 12 is not as far receded as point 11.

Spaces of higher dimensions are obviously no longer visible to the human eye, but can be adequately treated mathematically. They have been studied by statistical methods (e.g., principal component analysis, cluster analysis, kth nearest neighbor analysis, logistic regression analysis).[49] As a result of this investigation, functions could be developed that use the seven previously mentioned basic parameters to calculate the reactivity index of any bond within a molecule.

Consequently, the simulation of a reaction in a forward search is strongly dependent on the quality of the reactivity function, which undergoes steady improvement by inclusion of new experimental data in the reactivity space.

Philosophically, it is correct to try to develop functions that in one way or the other model the reactive behavior of a molecule. This approach makes a computer program autodeductive, through which predictions of unprecedented reaction products and reaction mechanisms are made feasible.

V. OTHER APPROACHES BASED ON MECHANISTIC STEPS

Other approaches use elementary mechanistic principles to describe and, thus, generate chemical reactions. They mimic the traditional mechanistic reasoning of the synthetic organic chemist by inclusion of very few, essential basic mechanisms.

SYNGEN[50,51] is an SDS utilizing half-reactions defined by a change in functionality from substrate to product for each of the two linking synthons. The half-reaction has an oxidoreductive character, and the unions of fragments must combine two half-reactions of opposite polarity. The two main strategies of the SYNGEN procedure are (1) the dissection of the molecular skeleton into approximately equal halves (selection of strategic bonds), which leads to a synthesis tree of minimal complexity (the minimal spanning tree of a retrosynthesis), and (2) the generation of the necessary functionality on the skeleton(s) to afford subsequent synthesis reactions. SYNGEN does not use a library of reactions, but generates each reaction from reaction mechanism theory.

Another very successful approach to computer-assisted modeling of reactions is the CAMEO program.[21,52] It predicts the products of organic reactions from given educts under known conditions. CAMEO operates by reproducing the elementary mechanistic reaction steps familiar to every organic chemist. It can be used for retrosyntheses as well as in a forward search manner. Its aim is to give an in-depth analysis of the feasibility of a certain individual reaction rather than to generate complete synthesis trees. It has the advantage of accounting for by-products of a reaction.

Some mechanisms included in the program are halogen-metal exchange, substitution, addition, elimination, base-catalyzed and organometallic reactions, nucleophilic and electrophilic processes involving organosilicon compounds, thermal pericyclic reactions, electrophilic aromatic substitution, and carbonium ion reactions. It must be stressed again that these are not huge collections of thousands of individual, published, and known reactions, but just their basic templates (the same templates we use when writing a reaction mechanism on a piece of paper). CAMEO has shown a high performance for the simulation of base-catalyzed reactions. Hundreds of processes can be represented in terms of chains of a few basic mechanistic steps.

A. SYNTHESIS DESIGN AND REACTION PREDICTION: ARTIFICIAL INTELLIGENCE, EXPERT SYSTEMS, OR ...?

An increasing number of more or less fully professional computer chemists describe such programs as examples of AI or of expert systems (ES). This tendency occurs even if such programs are not obedient to some criteria which should classify programs as belonging to the AI type, to the ES type, or just as programs.

It seems that in recent times the terms "artificial intelligence" and "expert system" attach some magical power to software packages, and the author in all honesty will not exclude himself from his responsibilities of being prey to this "attraction".

Unfortunately, it seems very problematic, even for professionals, to agree on final definitions of AI and ES. It was stated correctly that no rigorous definition is readily available, but that AI must have something to do with "... construction of a mechanizable logic of commonsense reasoning". As we are computer chemists who write programs that somehow try to model chemical events, and since what we write comes from our knowledge, we can reasonably assume that what we program obeys commonsense chemical reasoning. But this is what we program. The computer so far does not draw its own conclusions with regard to new rules, meaning that it cannot yet program itself and introduce additional rules inferred totally from its own considerations. The program examples presented in the dedicated chapters are all dealing, in one way or another, with information provided by man. A structure elucidation system cannot find spin-splitting rules on its own, and if no indication is given explicitly no program will generate pentacoordinated carbon atoms in charged organic molecules. The separation line between AI programs in the strict sense and computer chemistry programs is fluctuating. The author prefers to address these programs with the name "autodeductive system" (AD). An AD system will not extract new rules, but by itself (auto-) finds (-deduce) solutions to a given problem. Its "intelligence" will then be confined to its capability of evaluating the most favorable (or most probable) solution from a set of possible ones. This is quite an achievement in organic chemistry!

As ES is characterized by some knowledge base, which stores rules, interfaced with a rule interpreter; the rules are frequently empirical in nature, like rules-of-thumb. The computer combines the effects of these rules and drives at a "logical" conclusion, given some starting conditions. Thus, an ES can be regarded as that part of an AI system left after the requirement of working out its own rules has been stripped from the AI.

Perhaps computer chemistry systems should lie closer to the area of ES, as they are constructed on a "distilled knowledge of organic chemistry", or, more simply, because they are made by experts!

REFERENCES

1. **Vernin, G. and Chanon, M.**, Eds., *Computer Aids to Chemistry*, Ellis Horwood, Chichester, U.K., 1986.
2. **Wipke, W. T., Heller, S. R., Feldmann, R. J., and Hyde, E.**, Eds., *Computer Representation and Manipulation of Chemical Information*, Interscience, New York, 1974.
3. **Wipke, W. T. and Howe, W. J.**, Eds., *Computer-Assisted Organic Synthesis*, ACS Symp. Ser., Vol. 61, American Chemical Society, Washington, D.C., 1977.

4. **Wipke, W. T., Braun, H., Smith, G., Choplin, F., and Sieber, W.,** SECS — simulation and evaluation of chemical synthesis: strategy and planning, in *Computer-Assisted Organic Synthesis*, ACS Symp. Ser., Vol. 61, Wipke, W. T. and Howe, W. J., Eds., American Chemical Society, Washington, D.C., 1977, 97.

5. **Wipke, W.R., Ouchi, G. I., and Krishnan, S.,** SECS: an application of artificial intelligence techniques, *Artif. Intell.*, 9, 173, 1978.

6. **Wipke, W. T. and Dolata, D.,** A multivalued logic predicate calculus approach to synthesis planning, in *Artificial Intelligence Applications in Chemistry*, ACS Symp. Ser., Vol. 306, Pierce, T. H. and Hohne, B. A., Eds., American Chemical Society, Washington, D.C., 1986, 189.

7. **Pensak, D. and Corey, E. J.,** LHASA — logic and heuristics applied to synthetic analysis, in *Computer-Assisted Organic Synthesis*, ACS Symp. Ser. Vol. 61, Wipke, W. T. and Howe, W. J., Eds., American Chemical Society, Washington, D.C., 1977, 1.

8. **Corey, E. J., Johnson, A. P., and Long, A. K.,** Computer-assisted synthetic analysis. Techniques for efficient long-range retrosynthetic searches applied to the Robinson annulation process, *J. Org. Chem.*, 45, 2051, 1980.

9. **Hippe, Z.,** Self-adapting computer program for designing organic syntheses, *Anal. Chim. Acta*, 133, 677, 1981.

10. **Gelernter, H., Sridharan, N. S., Hart, A. J., Yen, S. C., Fowler, F. W., and Shue, H.,** The discovery of organic synthesis routes by computer, *Top. Curr. Chem.*, 41, 113, 1973.

11. **Agarwal, K. K., Larsen, D. L., and Gelernter, H.,** Applications of chemical transformations in SYNCHEM2, a computer program for organic synthesis route discovery, *Comput. Chem.*, 2, 75, 1978.

12. **Corey, E. J. and Wipke, W. T.,** Computer-assisted design of complex organic syntheses, *Science*, 166, 178, 1969.

13. **Corey, E. J. and Jorgensen, W.,** Computer-assisted synthetic analysis. Synthetic strategies based on appendages and the use of reconnective transforms, *J. Am. Chem. Soc.*, 98, 189, 1981.

14. **Hendrickson, J. B.,** Systematic synthesis design. VI. Yield Analysis and Convergency, *J. Am. Chem. Soc.*, 99, 5439, 1977.

15. **Corey, E. J., Long, A. K., Mulzer, J., Orf, H. W., Johnson, A. P., and Hewett, A. P.,** Computer-assisted synthetic analysis. Long-range search procedures for antithetic simplification of complex targets by application of the halolactonization transfrom, *J. Chem. Inf. Comput. Sci.*, 20, 221, 1980.

16. **Hendrickson, J. B. and Braun-Keller, E.,** Systematic synthesis design. Generation of reaction sequences, *J. Comput. Chem.*, 1, 323, 1980.

17. **Choplin, F., Laurenco, C., Marc, R., Kaufmann, G., and Wipke, W. T.,** Synthese assistee par ordinateur en chimie des composees organophosphores, *Nouveau J. Chim.*, 2, 285, 1978.

18. **Zimmer, M. H., Choplin, F., Bonnet, P., and Kaufmann, G.,** Automatic strategy in computer design of synthesis: an example in organophosphorous chemistry, *J. Chem. Inf. Comput. Sci.*, 4, 235, 1979.

19. **Choplin, F., Bonnet, P., Zimmer, M. H., and Kaufmann, G.,** Interactive strategy in computer design of synthesis, *Nouveau J. Chim.*, 3, 223, 1979.

20. **Gelernter, H., Bhagwat, S. S., Larsen, D. L., and Miller, G. A.,** Knowledge-base enhancement via training sequence: the education of SYNCHEM-2, in *Computer-Applications in Chemistry*, Anal. Chem. Symp. Ser., Vol. 15, Heller, S. R. and Potenzone, R., Eds., Elsevier, Amsterdam, 1983, 35.

21. **Dugundji, J. and Ugi, I.,** Algebraic model of constitutional chemistry as a basis for chemical computer programs, *Top. Curr. Chem.*, 39, 19, 1973.

22. **Schubert, W. and Ugi, I.,** Darstellung chemischer Strukturen fuer die computerunterstuetzte deduktive Loesung chemischer Probleme, *Chimia*, 6, 183, 1979.

23. **Jochum, C., Gasteiger, J., Ugi, I., and Dugundji, J.,** The principle of minimal chemical distance and the principle of minimum chemical structure change, *Z. Naturforsch. Teil B*, 37, 1205, 1982.

24. **Jochum, C.,** Algorithmen zur Auswertung konstitutioneller Information organisch-chemischer Strukturen, Ph.D. thesis, Technical University Munich, Munich, Federal Republic of Germany, 1979.

25. **Burkard, R. E.,** Heuristische Verfahren zuer Loesung quadratischer Zuordnungsprobleme, *Z. Oper. Res.*, 19, 183, 1975.

26. **Weise, A.,** Ableitung organisch-chemischer Reaktionen mit dem Simulationsprogramm AHMOS, *Z. Chem.*, 15, 333, 1975.

27. **Ugi, I., Bauer, J., Brandt, J., Dugundji, J., Frank, R., Friedrich, J., Scholley, A. V., and Schubert, W.,** Mathematical model of constitutional chemistry and system for computer programs for deductive solution of chemical programs for deductive solution on chemical problems, in *Data Processing in Chemistry*, Studies in Physical and Theoretical Chemistry, Vol. 16, Hippe, Z., Ed., Elsevier, Amsterdam, 1981, 219.

28. **Bures, M., Roos-Kozel, B., and Jorgensen, W. L.,** Computer-assisted mechanistic evaluation of organic reactions. XI. Electrophilic aromatic substitution, *J. Org. Chem.*, 50, 4490, 1985.

29. **Bauer, J., Herges, R., Fontain, E., and Ugi, I.,** IGOR and computer-assisted innovation in chemistry, *Chimia*, 2, 43, 1985.

30. **Gasteiger, J., Jochum, C., Marsili, M., and Thoma, J.,** Das Syntheseplanungsprogramm EROS, *MATCH,* 6, 177, 1979.

31. **Marsili, M., Gasteiger, J., and Carter, R. E.,** Computer-assisted organic chemistry: an introduction into the EROS system, *Chim. Oggi,* 9, 11, 1984.

32. **Gasteiger, J., Hutchings, M., Christoph, B., Gann, L., Hiller, C., Loew, P., Marsili, M., Saller, H., and Yuki, K.,** A new treatment of chemical reactivity: development of EROS, an expert system for reaction prediction and synthesis design, *Top. Curr. Chem.,* 137, 21, 1987.

33. **Gasteiger, J.,** An algorithm for estimating heats of reaction, *Comput. Chem.,* 2, 85, 1978.

34. **Gasteiger, J., Jakob, P., and Strauss, U.,** Critical evaluation of additivity schemes for estimating heats of atomization, *Tetrahedron,* 35, 139, 1979.

35. **Weissermel, K. and Arpe, H. J.,** *Industrielle Organische Chemie,* Verlag Chemie, Weinheim, Federal Republic of Germany, 1976, 286.

36. **Weissermel, K. and Arpe, H. J.,** *Industrielle Organische Chemie,* Verlag Chemie, Weinheim, Federal Republic of Germany, 1976, 285.

37. **Gasteiger, J. and Dammer, O.,** Automatic estimation of ring strain energies, *Tetrahedron,* 34, 2939, 1978.

38. **Gasteiger, J.,** A representation of pi-systems for efficient computer manipulation, *J. Chem. Inf. Comput. Sci.,* 19, 111, 1979.

39. **Gasteiger, J. and Marsili, M.,** Iterative partial equalization of orbital electronegativity — a rapid access to atomic charges, *Tetrahedron,* 36, 3219, 1989.

40. **Taft, R. W.,** Separation of polar, steric and resonance effects in reactivity, in *Steric Effects in Organic Chemistry,* Newman, M. S., Ed., John Wiley & Sons, New York, 1956, 556.

41. **Gasteiger, J., Marsili, M., and Paulus, B.,** Investigation into chemical reactivity and planning of chemical syntheses, in *Data Processing in Chemistry,* Studies in Physical and Theoretical Chemistry, Vol. 16, Hippe, Z., Ed., Elsevier, Amsterdam, 1981, 229.

42. **Marsili, M.,** Ladungsverteilungen und Reaktivität in der computer gesteuerten Reaktions-simulation, Ph.D. thesis, Technical University Munich, Munich, Federal Republic of Germany, 1980.

43. **Gasteiger, J., Jochum, C., Marsili, M., and Thoma, J.,** in Proc. 4th Symp. Use of Computers in Organic Chemistry, Stockholm, Sweden, 1981, 51.

44. **Gasteiger, J. and Hutchings, M.,** Quantification of effective polarizability. Applications to studies of X-ray photoelectron spectroscopy and alkylamine protonation, *J. Chem. Soc. Perkin Trans. 2,* p. 559, 1984.

45. **Gasteiger, J. and Hutchings, M.,** Quantitative models of gas-phase proton transfer reactions involving alcohols, ethers, and their thio analogs: correlation analyses based on residual electronegativity and effective polarizability, *J. Am. Chem. Soc.,* 106, 6489, 1984.

46. **Hutchings, M. and Gasteiger, J.,** A quantitative description of fundamental polar reaction types. Proton and hybride transfer reactions connecting alcohols and carbonyl compounds in the gas phase, *J. Chem. Soc. Perkin Trans. 2,* p. 447, 1986.

47. **Hutchings, M. and Gasteiger, J.,** Correlation analyses of aqueous phase acidities of alcohols and gem-diols, and of carbonyl hydration equilibria, using electronic and structural parameters, *J. Chem. Soc. Perkin Trans. 2,* p. 455, 1986.

48. **Gasteiger, J., Hutchings, M., Loew, P., and Saller, H.,** The acquisition and representation of knowledge for expert systems in organic chemistry, in *Artificial Intelligence Applications in Chemistry,* ACS Symp. Ser., Vol. 306, Pierce, T. H. and Hohne, B. A., Eds., American Chemical Society, Washington, D.C., 1986, 258.

49. **Gasteiger, J., Saller, H., and Loew, P.,** Elucidating chemical reactivity by pattern recognition methods, *Anal. Chim. Acta,* 191, 111, 1986.

50. **Hendrickson, J. B., Grier, D. L., and Toczko, A. G.,** A logic-based program for synthesis design, *J. Am. Chem. Soc.,* 107, 5228, 1985.

51. **Hendrickson, J. B.,** Approaching the logic of synthesis design, *Acc. Chem. Res.,* 19, 274, 1986.

52. **Salatin, T.D. and Jorgensen, W. L.,** Computer-assisted mechanistic evaluation of organic reactions. I. Overview, *J. Org. Chem.,* 45, 2043, 1980.

APPENDIX

WELCOME TO CONGEN, VERSION VI.
CONGEN IS A PROGRAM FOR COMPUTER-ASSISTED STRUCTURE ELUCIDATION
DEVELOPED WITH NIH SUPPORT BY THE DENDRAL GROUP AT STANFORD.
MAY I RECORD SESSION?: no
#define molform c 6 h 6
MOLECULAR FORMULA DEFINED
#generate
SUPERATOM
'COLLAPSED' FORMULA IS C 6 H 6
CONSTRAINT:

....
....
....
.

217 STRUCTURES WERE GENERATED
#draw at (1 217)

```
#1:

      C—C
     / | / |
  C—*—C
  | / | /
  C—C


#2:

  C———C
  | \   | \
  |  C—+—C
  | /   | /
  C———C


#3:

  C—C—C
  = |  =
  C—C—C

#4:

  C         C
  =\       /=
  C—C——C—C
```

```
#5:

     C——C
    / |    =
   /  |    =
 C——C   C
   \  |  /
    \ | /
      C

#6:

    C—C
  =     =
 C        C
   \     /
    C=C

#7:

    C———C
   / \   | \
 C———C—+—C
      \ | /
        C

#8:

       C—C
      / X |
 C==C—C—C
```

```
#9:

 C
 =\
 C—C—C
    \ | \
      C—C

#10:

     C—C
    / | /  \
 C—C   C
    \   =
      C

#11:

 C—C—C=C
 = | /
 C—C

#12:

       C=C
      /   |
 C=C—C=C

#13:

 C
 =\
 C—C==C—C=C
```

```
#14:

C-C-C
= | |
C-C=C

#15:

   C-C
  / | =
 C  |  C
  \ | /
   C=C

#16:

     C
    /|
   / |
 C--|C
 =  |  =
 C--C--C

#17:

 C         C
 |=       /=
 C-C--C-C

#18:

 C
 =\
 C-C--C=C=C

#19:

          C
         /=
 C#C-C--C-C

#20:

    C---C
   / \ /|
  C   C |
   \ / \|
    C---C
```

```
#21:

     C
   /|\
  / | \
 C  C  C
 =  |\  |
 =|  \|
   C--C

#22:

   C-C
  = X \
 C-C-C-C

#23:

    C
   / \
 C-C-\-C
  \|  \|
   C---C

#24:

 C---C
 |\ /|
 C C |
 |  \|
 C===C

#25:

   C-C
  =| |
 C C-C
  \|/
   C

#26:

     C
    / =
   C   C
  /|   =
 C-C-C

#27:

     C
    /|=
   C | C
  /| |/
 C-C-C
```

```
#28:

 C-C
 \|\
 C-C--C#C

#29:

 C     C
 =\ / \
 C-C   C
    \ =
     C

#30:

      C
     / \
    C   C
   /|   =
 C=C-C

#31:

   C=C
  /   \
 C     C
  \     =
   C=C

#32:

   C-C
  =| \
 C |  C
  \|   =
   C-C

#33:

    C
   / \
  C   C-C#C
  = /
   C

#34:

    C
   /|\
 C-+-C--C==C
  \|/
   C
```

#35:

```
         C
        / \
  C=C-C   C
        \ =
         C
```

#36:

```
         C
        / =
  C=C-C   C
        \ =
         C
```

#37:

```
            C
           /|\
          / | \
         /  |  C
        /   |  =
  C=C--C---C
```

#38:

```
  C=C-C=C-C#C
```

#39:

```
  C-C-C
  :X X
  C-C-C
```

#40:

```
      /C
     / |
  C===C
  |   | \
  C--C==C
```

#41:

```
      C-C==C
     /|:/
  C==C-C
```

#42:

```
  C=C=C-C=C=C
```

#43:

```
        C
       / \
  C=C   C
     \ /=
      C-C
```

#44:

```
        C
       / \
  C-C--C
 =/      =
 C        C
```

#45:

```
  C--C--C
  \/ |  =
  /\ |  =
  C--C--C
```

#46:

```
   C=C
   | |
   C-C
  =   =
 C     C
```

#47:

```
         C
        /
       /  |
  C--C   C
   \  ==  |
    \ ==  |
     C   C
```

#48:

```
     C-C
    /  =
  C=C-C=C
```

#49:

```
     C-C
    /|:/
  C=C  C
     | =
     C
```

#50:

```
    C C
   /:X:|
  C-C C
     \=
      C
```

#51:

```
  C
  |\
  C=C--C=C=C
```

#52:

```
       C
      / \
    C---C-C#C
  =
  C
```

#53:

```
  C=C=C-C--#C
```

#54:

```
          C
         /:|
  C#C-C--C=C
```

#55:

```
  C-C---C
  \:\ / \
   C-C---C
```

#56:

```
      C
     / \
  C C   C
   \:\ =
    C-C
```

#57:

```
  C-C=C
  | | |
  C=C-C
```

```
#58:

C           C
|  =      / |
C—C——C=C
```

```
#59:

    C———C
   / \ / |\
  C   X  | C
   \ / \ |/
    C———C
```

```
#60:

    C=C
   / |  \
  C  |   C
   \ |  /
    C=C
```

```
#61:

C=C—C=C
|  | |/
C—C
```

```
#62:

C—C—C=C
|/ | /
C—C
```

```
#63:

    C—C
   /|X|
  C—C C
     \|
      C
```

```
#64:

      C
     / |
  C—C  |
   \|\ |
    C C |
     \=
      C
```

```
#65:

      C=C
     / |  \
C==C—+—C
     \  /
      C
```

```
#66:

      C
     / \
  C=C   C
   \ / |
    C=C
```

```
#67:

    C
   /|\
  C | C
   \| |\
    C=C—C
```

```
#68:

  C—C—C
 / ≠ /
C—C—C
```

```
#69

      C
     /=\
    C—=—C
   /  =  |
  C———C—C
```

```
#70:

        C
       /|\
C=C==C | C
       \|/
        C
```

```
#71:

    C
   /=\
  C = C—C
   \= |/
    C—C
```

```
#72:

        C
       /  =
C=C=C   C
     \  /
      C
```

```
#73:

    C—C
   /=   =
  C =   C
   \=  /
    C—C
```

```
#74:

    C=C
   /    \
  C      C
   \    /
    C#C
```

```
#75:

  C C
 =|X|
C—C C
   \|
    C
```

```
#76:

      C
     / |
C—C—C
 =| |
  C—C
```

```
#77:

        C
       / |
C#C——C—C
       \ |
        C
```

```
#78:

  C———C
 / \   |
C   C  |
 \  =  \|
  C———C
```

```
#79:

     C—C
    / |  \
   C  |   C
    \ |  =
     C=C


#80:

    C---C
   / \   |
  C   C  |
   \ / \ |
    C===C


#81:

           C
          / |
         /  |
   C==C  C  |
    =  \/ | |
    =  /\ | |
      C   C


#82:

           C
          / |
         /  |
   C--C  C  |
    #  \/ | |
    #  /\ | |
      C   C


#83:

    C—C
   /    \
  C  |   C
   # | /
    C—C


#84:

      C
     /  =
   C---C—C=C
   =
   C
```

```
#85:

         C
        / |
  C=C--C—C
        \ =
         C


#86:

       C
       =
       C
      / \
     C   C
     #     =
    C       C


#87:

      C
     / | =
    C  |  C--C=C
     \ | /
       C


#88:

      C
     / \
    C   C—C=C
    = =
     C


#89:

      C
     /=\
    C = C--C=C
     \=/
      C


#90:

  C=C=C=C—C=C


#91:

  C=C—C#C—C=C


#92:

       C
      #  \
    C     C—C=C
     \   /
      C
```

```
#93:

         C
        / =
  C=C—C--C=C


#94:

         C
        / #
  C=C—C--C—C


#95:

      C   C
     = \ / |
    C   C—C
     = /
      C


#96:

  C—C----C
  ==      |
   =      |
  ==      |
  C—C--C


#97:

  C—C==C—C#C
  | |/
  C


#98:

  C#C—C—C—C#C


#99:

  C—C--C
  | X |  |
  C—C--C


#100:

  C—C=C
  | | |
  C—C=C


#101:

         C
        / | =
   C   / |  C
   |\ / | |/
   C—C---C
```

#102:

```
C—C—C
|  =  =
C—C—C
```

#103:

```
C           C
| \       / =
C—C==C—C
```

#104:

```
     C—C
    =  |    \
 C     |     C
    \  |    /
     C=C
```

#105:

```
        C
      / \
 C—C     C
  =| \  /
    C—C
```

#106:

```
     C
   / \
 C     C—C#C
   \ =
     C
```

#107:

```
     C——C
    /  |   |
   /   |   |
 C==C   C
   \   |  /
    \  | /
     C
```

#108:

```
     C—C
    /     =
 C       C
   \     =
    C=C
```

#109:

```
        C———C
      / =     |
     /   =    C
 C——C   /
   \   |  /
    \  | /
     C
```

#110:

```
C—C=C
|  |   =
C—C—C
```

#111:

```
C—C—C
|  |  #
C—C—C
```

#112:

```
     C—C
    /     =
 C       C
   \     /
    C#C
```

#113:

```
 C           C
 | \       / =
 C—C——C=C
```

#114:

```
 C           C
 | \       / #
 C—C——C=C
```

#115:

```
        C—C
      /    #
 C=C—C—C
```

#116:

```
 C—C—C
 \ | \ |
  C—C
   \ |
    C
```

#117:

```
     C       C
   / | =  /
  /   |  /  ≠  |
 /    | /   = |
C———C   C
       \    /
        C
```

#118:

```
        C
      / \
 C=C     C
   \   = |
    C—C
```

#119:

```
 C——C
  \ / | \
  /\  |  \
 C——C——C==C
```

#120:

```
 C
   =
     C—C=C
     | /
     C
       =
     C
```

#121:

```
        C
      / \
     C———C=C=C
   =
 C
```

#122:

```
        C
      / \
 C—C     C
  \ | \ /
   C=C
```

#123:

```
 C—C
 |   = \
 C—C—C=C
```

#124:

```
C=C—C—C
       \   ¦
        C#C
```

#125:

```
          C
        /=\
   C   /  = C
   ¦\ /  =/
   C—C———C
```

#126:

```
     C   C
   #  \ /¦
   C   C—C
    \ /
     C
```

#127:

```
   C—C————C
    \=      ¦
     ≠      ¦
     =\     ¦
      C=C——C
```

#128:

```
   C—C==C=C=C
   ¦/
   C
```

#129:

```
   C   C
   #\ / \
   C—C   C
      \ /
       C
```

#130:

```
     C—C
   = ¦    \
   C ¦     C
   = ¦    /
     C—C
```

#131:

```
      C—C
    / ¦=
   C—C C
     =/
     C
```

#132:

```
     C—C
     =   =
   C—C—C=C
```

#133:

```
             C\\\C
            /  --\
           /  --¦  \
          ≠-    ¦    \
       ———C     ¦\
   C——     \    ¦  \
            \       C
             \     =
              \   =
               \ =
                C
```

#134:

```
         C
       = \
      C———C—C#C
     /
   C
```

#135:

```
      C
    /¦\
   C—C—C——C
    \¦/
     C
```

#136:

```
   C===C
    \  ¦ \
     C-+-C——C
    /  ¦ /
       C
```

#137:

```
         C
       = \
   C—C   C
    \  ¦ ¦
     \ ¦  C
      \ ¦ =
        C
```

#138:

```
        C—C
      = X ¦
   C——C—C—C
```

#139:

```
      C
      ¦=
      ¦  C
      ¦ / \
      C   C
     / \ =
    C   C
```

#140:

```
   C        C
    \      /=
     C———C—C
     = /
       C
```

#141:

```
      C
    = \
   C———C——C#C
        ¦
        ¦
        C
```

#142:

```
   C—C
   = =\
   C—C—C—C
```

#143:

```
   C
   =\
   C—C==C=C—C
```

#144:

```
C-C=C-C
=  | /
C-C
```

#145:

```
     C=C
    /   |
C-C=C=C
```

#146:

```
     C=C
    /   |
C-C-C#C
```

#147:

```
C
=\
C-C--C#C-C
```

#148:

```
        C
       / =
C-C=C    C
       \ =
        C
```

#149:

```
      C===C
     / |  /
C--C-+-C
     | =
     C
```

#150:

```
C-C=C=C-C#C
```

#151:

```
     C
     |
     C
    / \
   C   C
   #     #
   C       C
```

#152:

```
    C-C
   / |X|
C--C-C-C
```

#153:

```
C=C
| | |\
C=C-C-C
```

#154:

```
         C
        /|\
C-C==C | C
        \|=
         C
```

#155:

```
    C=C
    | | |\
C-C-C=C
```

#156:

```
C-C-C-C
=/ | /
C-C
```

#157:

```
       C
      / =
   C---C-C#C
  /
 C
```

#158:

```
C--C-C-C
    \ X=
     C-C
```

#159:

```
    C=C
   /  =
C-C-C=C
```

#160:

```
       C=C
      / X|
C--C-C-C
```

#161:

```
C-C
=  |\
C-C-C-C
```

#162:

```
C-C-C-C
#  | /
C-C
```

#163:

```
C
=\
C=C--C=C-C
```

#164:

```
C
#\
C-C--C=C-C
```

#165:

```
    C-C
   /  #
C-C=C-C
```

#166:

```
C-C#C-C=C=C
```

#167:

```
C-C#C-C-C#C
```

#168:

```
C       C
 \     /#
  C--C-C
 =
C
```

```
#169:

C
 \
  C---C=C
  |\ /
  C=C
```

```
#170:

      C
   =   \
  C---C=C=C
 /
C
```

```
#171:

C---C
\ /|\
 C-+-C--C
\ |/
   C
```

```
#172:

      C
   =   \
 C-C    C
   \ =|
    C-C
```

```
#173:

    C
   /=\
  C-=-C--C
   \=
    C
    \|
     C
```

```
#174:

      C-C==C
     /|/
  C--C=C
```

```
#175:

   C=C
   =
   C-C
  /    =
 C      C
```

```
#176:

    C=C
   /|/
C--C-C
   |/
   C
```

```
#177:

    C=C
   /  =
C-C-C
   |/
   C
```

```
#178:

C       C
 \     ==
  C--C-C
 =
C
```

```
#179:

    C-C
   =|  |
C--C-+-C
       |
       =
       C
```

```
#180:

    C
   /   =
 C-C    C
   =  /|
    C-C
```

```
#181:

    C
   / \
 C-C    C
   =  /=
    C-C
```

```
#182:

    C
    |
    C
   =  \
  C    C
  =      #
  C      C
```

```
#183:

     C
    / \
   C===C-C#C
  /
 C
```

```
#184:

       C
      /=\
 C-C==C  =  C
      \=/
       C
```

```
#185:

      C
     / #
 C-C=C    C
     \ /
      C
```

```
#186:

 C-C-C-C
 |/|/
 C=C
```

```
#187:

     C-C==C
    /=/
 C--C-C
```

```
#188:

    C#C
    |  |
    C-C
   /     =
  C      C
```

```
#189:

       C=C
      /|   =
 C--C-+-C
      |/
      C
```

```
#190:

 C-C=C=C=C=C
```

#191:

```
C
| \
|  \
C=C--C#C—C
```

#192:

```
      C
    = | \
    C-+-C--C
    /  | /
  C---C
```

#193:

```
      C—C
    =   | |
  C—C—C#C
```

#194:

```
C—C—C—C
| / | =
C—C
```

#195:

```
      C
    = | \
    C-+-C--C
    \ |  |
     C  |
      \ |
       C
```

#196:

```
     C—C
    / \ |
   C   C
   # / \
    C   C
```

#197:

```
      C—C
     / | /
   C—C C
    ==
     C
```

#198:

```
C—C=C—C
|  | |/
C=C
```

#199:

```
     C—C
    /   =
  C—C=C=C
```

#200:

```
C           C
 \        / #
  C---C—C
   \ /
    C
```

#201:

```
       C
      / \
  C—C   C
     \ /=
      C=C
```

#202:

```
        C
        |
        |
  C---C--C=C
   #  /
      C
```

#203:

```
C
 \
  C—C—C=C
  ==
   C
```

#204:

C—C—C#C—C#C

#205:

```
     C
    / | \
  C-+-C--C--C
  = | /
    C
```

#206:

```
         C
        / #
  C—C—C   C
      = /
        C
```

#207:

```
         C
        / | =
  C—C--C | C
        = | /
          C
```

#208:

```
      = =
  C—C—C   C
       \ =
        C
```

#209:

```
C
# \
 C—C==C—C—C
```

#210:

```
          C
         / =
        /   C
       /    =
  C—C--C---C
```

#211:

```
  C     C
   \   /
    C—C
    | X |
    C=C
```

#212:

```
          C
         / | =
  C     / | C
   \   / |  =
    C---C
   /
  C
```

#213:

```
   C#C
   | |
   C=C
  /     \
 C       C
```

#214:

```
C           C
 \         /#
   C==C—C
  /
 C
```

#215:

```
        C
       /|=
C——C   |  C——C
      =|/
        C
```

#216:

```
   C=C
   = =
   C—C
  /     \
 C       C
```

#217:

C—C#C—C#C—C

INDEX

Milton Keynes UK
Ingram Content Group UK Ltd.
UKHW051934141024
449569UK00027B/1486

9 781138 557888